PRE-TREATMENT METHODS OF LIGNOCELLULOSIC BIOMASS FOR BIOFUEL PRODUCTION

PRE-TREATMENT METHODS OF LIGNOCELLULOSIC BIOMASS FOR BIOFUEL PRODUCTION

Shyamal Roy

CRC Press
Taylor & Francis Group
Boca Raton London New York

CRC Press is an imprint of the
Taylor & Francis Group, an **informa** business

First edition published 2022
by CRC Press
6000 Broken Sound Parkway NW, Suite 300, Boca Raton, FL
33487-2742

and by CRC Press
2 Park Square, Milton Park, Abingdon, Oxon, OX14 4RN

© 2022 Shyamal Roy

CRC Press is an imprint of Taylor & Francis Group, LLC

ISBN: 978-1-032-06692-9 (hbk)
ISBN: 978-1-032-06693-6 (pbk)
ISBN: 978-1-003-20341-4 (ebk)

DOI: 10.1201/9781003203414

Typeset in Times
by MPS Limited, Dehradun

TABLE OF CONTENTS

LIST OF FIGURE

LIST OF TABLES

AUTHOR

Dr. Shyamal Roy completed BE & ME in Chemical Engineering from Jadavpur University, Kolkata, India and PhD in Chemical Engineering from Indian Institute of Technology Delhi, New Delhi, India. He completed postdoctoral research from Washington University in St. Louis, MO, USA and Rutgers University, NJ, USA. He also worked as a postdoctoral fellow in Research & Development Centre, Bharat Petroleum Corporation Ltd. He was awarded prestigious Raman Fellowship by UGC, Govt. of India in the year of 2015. Now he is a faculty member at Jadavpur University, Kolkata, India.

1

INTRODUCTION

1.1 INTRODUCTION

The development of sustainable energy systems based on renewable lignocellulosic biomass feed stocks is now a global endeavor. The shifting of the crude oil-based refinery to biomass-based biorefinery has enticed compelling scientific interest which focuses on the development of cellulosic ethanol as an alternative transportation fuel to fossil fuels. It is needless to say that of late biofuel has drawn attention as a renewable energy resource that may help cope with rising fuel prices, address environmental concerns associated with greenhouse gas emissions, and create new earnings and employment opportunities for the people in rural places around the globe. Release of CO_2 increases through the combustion of fossil fuel, and it is a major concern of global warming. The United States having only 4.5% of the world's population utilizes 25% of global energy consumption and responsible for 25% of global CO_2 emissions.[1] Therefore, conversion of plentiful lignocellulosic biomass to biofuel as transportation fuels will be a feasible alternative for boosting energy security and abating greenhouse gas emissions.[2,3] Unlike fossil fuels that originate from plants that grew millions of years ago, lignocellulosic biofuels are synthesized from plants grown today. Lignocellulosic biofuel has the great potential to reduce greenhouse gas emissions by 85%.[4] Lignocellulosic biomasses such as forest products (hardwood and softwood), dedicated crops (switch grass, salix) and agricultural residues (wheat straw, sugarcane bagasse, corn stover, etc.) are sustainable energy resources.[5,6] Lignin, cellulose, hemicellulose and pectin contribute ca. 90% of the dry weight of most plant biomasses.[1] The lignin acts as a protective wall that resists plant cell destruction by bacteria and fungi for conversion to fuel. It is essential to break down the hemicellulose and cellulose into their respective monomers (sugars) for producing fuel from lignocellulosic biomass.[7] Some physicochemical, structural, and compositional features resist the digestibility of cellulose existing in lignocellulosic biomass. Pretreatment of lignocellulosic biomass is

DOI: 10.1201/9781003203414-1

essential for exposing the cellulose in the plant fibers prior to conversion of lignocellulosic biomass to biofuel using the digestion technique. Various pretreatment methods are used such as chemical, biological, steam explosion, ammonia fiber explosion, and extractive ammonia for modification of the structure of lignocellulosic materials in order to augment the accessibility of the cellulose towards enzyme during hydrolysis.[8] The pretreatment process is used to disintegrate the lignin structure and deconstruct the cellulose structure so that the cellulose hydrolysis rate increases with favoring the higher accessibility of enzymes or acids.[9–13] Pretreatment is a salient technique for biofuel production from lignocellulosic biomass and our effort to survey the recent advances in pretreatment methods for the economic design of the sustainable biofuel production process.

1.2 STRUCTURE OF LIGNOCELLULOSIC BIOMASS

In general, the lignocellulosic biomass consists of lignin, hemicelluloses, cellulose, polyoses, and trace amounts of pectin, protein, extractives (soluble nonstructural materials such as nonstructural sugars, nitrogenous material, chlorophyll, and waxes), and ash.[14] Biomass cell wall consists of lignin, hemicelluloses, and cellulose. The structural segment of plant cell walls consists of polysaccharides. The key factor of economic bio refining is untwisting the complex polymeric structures. Cellulose microfibrils made of a crystalline structure of thousands of strands and thousands of sugar molecules form each strand. These microfibrils are enveloped in hemicelluloses and lignin, which prevents the cellulose from microbial attack (as shown in Figure 1.1). Hemicelluloses are relatively easy to break down using pretreatment process. It also disrupts the hemicelluloses or lignin enveloped around the cellulose, which results in the accessibility of cellulose for further hydrolysis. The hydrolysis of the lignocellulosic biomass will liberate C_6 fermentable sugars will be carried out via enzymatic hydrolysis.[15] It is reported that low ash content in lignocellulosic biomass enhances the enzymatic hydrolysis process.[16] The composition of the different constituents can vary from one plant species to another. For example, hard wood has greater amounts of cellulose, whereas wheat straw and leaves have more hemicellulose (as described in Table 1.1). Interestingly the ratios between various components within a single plant differ with age, stage of growth, and other conditions.[18] The main structural constituent in plant cell walls is

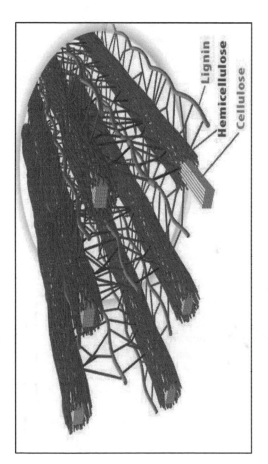

Figure 1.1 Cellulose, lignin, and hemicellulose are arranged in a specific way inside the cell walls of lignocellulcse. Adapted from U.S. Department of Energy Genome Programs image gallery (http://genomics.energy.gov).

TABLE 1.1

Cellulose, Hemicellulose, and Lignin Contents in Common Agricultural Residues and Wastes

Lignocellulosic Material	Cellulose (%)	Hemicellulose (%)	Lignin (%)
Hardwood stems	40–55	24–40	18–25
Softwood stems	45–50	25–35	25–35
Nutshells	25–30	25–30	30–40
Corn cobs	45	35	15
Grasses	25–40	35–50	10–30
Paper	85–99	0	0–15
Wheat straw	30	50	15
Sorted refuse	60	20	20
Leaves	15–20	80–85	0
Cottonseed hairs	80–95	5–20	0
Newspaper	40–55	25–40	18–30
Waste papers from chemical pulps	60–70	10–20	5–10
Primary wastewater solids	8–15	-	-
Solid cattle manure	1.6–4.7	1.4–3.3	2.7–5.7
Coastal bermudagrass	25	35.7	6.4
Switchgrass	45	31.4	12
Swine waste	6.0	28	NA

cellulose, and this biopolymer comprises D-glucose subunits linked to each other by β-(1,4)-glycosidic bonds.[19] The hydrogen and van der Waals bonds facilitate a long chain of cellulose and oblige to pack into microfibrils. Both crystalline and amorphous forms of cellulose exist in the biomass, and a very less percentage of incoherent cellulose chains contribute amorphous cellulose which is more susceptible to enzymatic degradation.[20] Hemicellulose possesses small lateral branch chains containing various sugar molecules, whereas cellulose does not have any branch chain containing sugar molecules. These monosaccharides are hexoses (such as glucose, mannose, and galactose), pentoses (such as xylose, rhamnose, and arabinose), and uronic acids (such as 4-o′methylglucuronic, D-glucuronic, and D-galacturonic acids). The foundation of hemicellulose is either a homopolymer or a hetero-polymer with small branches looped by β-(1,4)-glycosidic bonds and occasionally β-(1,3)-glycosidic bonds.[21] In comparison to cellulose, the polymers that exist in hemicelluloses do not conglomerate when they co-crystallize with cellulose chains and are comfortably hydrolysable. Lignin consists of cross-linked polymers of phenolic monomers and exists in the primary cell wall and provides structural support by resisting microbial attack.[18] Lignin contains three phenyl propionic alcohols such as sinapyl alcohol (syringyl alcohol), coniferyl alcohol (guaiacyl propanol), and coumaryl alcohol (p-hydroxyphenyl propanol), and phenolic monomers exist simultaneously linking alkyl aryl, alkyl alkyl, and aryl aryl ether bonds.[22] Higher content of lignin is found in softwoods, whereas herbaceous plants such as grasses possess the lower lignin content (as described in Table 1.1).

1.3 PRETREATMENT OF LIGNOCELLULOSIC BIOMASS

Pretreatment is a pivotal step in biochemical conversion process of lignocellulosic biomass into biofuel.[24] The purpose of the pretreatment process is to disrupt and eliminate hemicellulose and lignin, decrystallize the cellulose, and augment the active surface area and porosity of the lignocellulosic biomass materials. The recent advances with brief advantages and disadvantages of the following pretreatment techniques for conversion of lignocellulosic biomass to fuels and chemicals have been reviewed.

2

PHYSICAL PRETREATMENTS

2.1 MECHANICAL COMMINUTION

Biomass materials are comminuted by various chipping, grinding, and milling techniques. Biomass particle size is reduced using mechanical comminution for easy handling and increasing surface/volume ratio. The specific size of the biomass materials uses ca. 10–30 mm using chipping and ca. 0.2–2 mm using milling or grinding.[17] It was reported that the higher digestibility of spruce and aspen chips was observed using reduced cellulose crystallinity, and vibratory ball milling was found to be more effective than ordinary ball milling.[25] The power requirement for mechanical communication of lignocellulosic biomass depends on the final particle size and biomass characteristics.[26] It was studied that the energy input for comminution can be reduced below 30 kWh per ton of biomass using the final particle size in the range of 3–6 mm, and typically the theoretical energy was lower than the energy consumption content available in the lignocellulosic biomass. Milling operation can be carried out before and after chemical pretreatment, but it was reported that milling operation after chemical pretreatment remarkably decreases i) consumption of energy, ii) cost of separation of solid from liquid as pretreated particles can be separated out effortlessly, iii) energy required for intense mixing of pretreatment slurries, iv) solid to liquid ratio, and v) creation of fermentation inhibitors.[27] Barkat and his coworkers[28] reviewed the mechanical pretreatment for lignocellulosic biomass materials, and they reported that mechanical fractionation is a potential step to enhance final carbohydrate output, appropriate particle sizes and densification, enzymatic accessibility, and bioconversion affectivity without the synthesis of toxic products.

Recently, Madison et al.[29] studied the acoustic and hydrodynamic cavitation as satisfactory mechanical pretreatments for lignocellulosic biomass compounds. Microcrystalline cellulose and lime-treated sugarcane bagasse were used for acoustic cavitation, but raw and lime-treated sugarcane bagasse was used for hydrodynamic cavitation.

 DOI: 10.1201/9781003203414-2

They reported that acoustic cavitation enhanced microcrystalline cellulose enzymatic digestibility by 37% in comparison to no acoustic cavitation treatment. On the contrary, there was no significant change in lime-treated sugarcane bagasse. Hydrodynamic cavitation improved the enzymatic digestibility of both raw and lime-treated sugarcane bagasse. Finally, 46% sugar yield was found from enzymatic digestion 3 day using cavitation pretreated bagasse followed by lime treatment.

2.2 PYROLYSIS OF LIGNOCELLULOSIC BIOMASS

The pyrolysis technique is utilized for the pretreatment of lignocellulosic biomass. In this technique, lignocellulosic biomass is degraded thermally in the absence of oxygen, and cellulose quickly degraded into gaseous products and leftover char using temperature greater than $300°C$.[30,31] It was studied that slow rate of decomposition and production of less volatile components was found at low temperature, and 80–85% conversion of cellulose to reducing sugars with 50% glucose was achieved from acidic hydrolysis under moderate condition (97°C, 1 (N) H_2SO_4, 2.5 h).[32] High-grade Fischer-Tropsch (FT) fuels are synthesized using biomass to liquids (BtL) route.[33] For conversion of lignocellulosic biomass to FT fuel various pretreatment processes such as pyrolysis, chipping, torrefaction, and pelletization are used. Undri et al.[34] studied the pyrolysis of wood pellets using a microwave oven in the presence of Fe or carbon. Three fractions such as gas, liquid also called bio-oil, and a solid called bio-char were collected. Cellulose pyrolysis products are found in the upper phase such as water, acetic acid, furans, carbohydrates, and their derivatives. Compounds from pyrolysis of lignin such as phenols and veratric acid are present in the bottom phase. Bartoli et al.[35] carried out the microwave-assisted pyrolysis of stump-roots and leaves from various residues of short rotation coppice of poplar clones, and they reported that various chemicals such as acetic acid, formic acid, acetic anhydride, furanes, and different phenols are synthesized, and among all bio-oils acetic acid concentration with 543.3 mg/mL is obtained.

2.3 CATALYTIC FAST PYROLYSIS

Catalytic fast pyrolysis (CFP) is an advanced technique for pretreatment of lignocellulosic biomass. In this method, biomass is charged with fluidizing gas and hot catalyst in a pyrolysis reactor for a short duration at high temperature (~500°C) and atmospheric pressure in the absence of air. Lignocellulosic biomass is transformed into condensable vapors and non-condensable gas, char, and coke.[36] The progressive vapors are transformed into less reactive compounds in the presence of catalyst than the vapors evolved by fast pyrolysis without catalyst.[37,38] It was reported that higher energy density biofuel can be synthesized from low energy density lignocellulosic biomass using fast pyrolysis technique.[39,40] The study is extensively carried out on fast pyrolysis, a single-step biofuel upgrading process, in which the hot vapors evolved from pyrolysis are contacted with a catalytic bed before condensation.[41–45] Zeolite catalysts such as HZSM-5 are used extensively for the synthesis of biofuel (hydrocarbon) from lignocellulosic biomass using CFP method.[46–50] But highly light gas synthesis and quick deactivation of catalyst due to coke formation are reported.[51,52] It was studied that hydrogen deficient nature of lignocellulosic biomass facilitates the highly light gas synthesis and catalyst deactivation,[53] and higher amount of catalyst is required to attain high conversions.[54–57]

Recently, it is reported that molybdenum trioxide (MoO_3) is an effective catalyst that produces olefinic and aromatic hydrocarbons with significant selectivities (~97%) from various biomass-derived oxygenates under mild conditions.[58–64] Budhi et al.[65] reported that CFP of pine using molybdenum supported on KIT-5 mesoporous silica and phenols, furans, and small amounts of aromatic were synthesized. Nolte et al.[66] investigated the CFP of cellulose, lignin, and corn stover using MoO_3 catalyst at low H_2 pressure and they found linear alkanes and aromatics. Karthick Murugappan et al.[67] investigated the CFP of pine involving supported MoO_3 catalysts at high temperature and low H_2 pressure. They reported that olefinic, aromatic, furans, and phenols are synthesized depending on the ratio of catalyst to lignocellulosic biomass, and onset of catalyst deactivation is observed when catalyst to biomass ratio is higher(~5).[67]

2.4 ELECTRON BEAM IRRADIATION

Irradiation technology (especially electron beam irradiation) is extensively employed to change the feature of polymeric materials. This kind of technique increases the span of application for the irradiated materials. Microstructural crystallinity of the materials is modified using the irradiation method. Irradiation technique facilitates the chain-breaking mechanism by depolymerizing the polymeric substances. The cellulose crystal of the lignocellulosic biomass can be decrystallized by irradiation to frail fibrils.[68] The main advantage of the irradiation method is that breaking of the glucosidal bonds in the cellulose molecular chains can be performed in the presence of lignin. This method requires higher energy and hence is highly expensive and difficult for industrial-scale application. Jin et al.[69] studied the pretreatment of rice straw using electron beam irradiation and they reported that ca. 52% glucose was produced by enzymatic hydrolysis of electron beam irradiated sample, whereas ca. 22% glucose was found from untreated rice straw. Pretreatment of lignocellulosic material using γ-rays irradiation was reported, and it was noticed that γ-rays irradiation facilitates the breaking of 1,4-glycosidic bonds, generates higher active surface, and decreases cellulose crystallinity.[70] This method does not require an extreme temperature, and the production of some undesirable inhibitory compounds during alkali or acid pretreatment can be minimized or escaped. Bak[71] studied the electron beam irradiation of rice straw, and it was revealed that significant enzymatic hydrolysis and fermentation yields were achieved. It was noticed that when water-soaked rice straw (solid:liquid ratio of 100%) was treated with electron beam irradiation doses of 1 MeV at 80 kGy, 0.12 mA, 70.4% (theoretical maximum) of the glucose yield was achieved after 120 h of hydrolysis. Moreover, after simultaneous saccharification and fermentation for 48 h, the ethanol concentration, production yield, and productivity were 9.3 g/L, 57.0% of the theoretical maximum, and 0.19 g/L h, respectively. Karthika et al.[72] reported the effect of electron beam irradiation on a hybrid grass variety studied as a biomass pretreatment method using electron beam irradiation doses of 0, 75, 150, and 250 kGy. About 79% of the final sugar yield was achieved using Trichoderma reesei ATCC 26921 cellulase and enzyme loadings 30 FPU/g of biomass for 48 h at irradiation doses of 250 kGy. Recently, Loow et al.[73] reviewed the energy irradiation pretreatment of lignocellulosic biomass and their review highlights the recent research on energy irradiation for pretreating biomass as well as the industrial applications of reducing sugars in biotechnological, chemical, and fuel sectors.

2.5 MICROWAVE PRETREATMENT

Microwave irradiation is extensively used in several fields because of its elevated heating efficiency and effortless operation. This technique is associated with the application of electromagnetic waves to induce heat by the oscillation of molecules upon microwave absorption. Crystalline structure of cellulose can be modified using microwave heating.[74] It favors depolymerization of lignin and hemicellulose substances and hence facilitates the enzymatic susceptibility of lignocellulosic materials providing high surface area.[75] It was reported that microwave pretreated rice straw showed better results towards enzymatic hydrolysis in the presence of water[75,76] and glycerine medium with a small volume of water.[77] It was studied that the same hydrolysis rate and reducing sugar yield were found using microwave pretreated rice straw and the raw straw separately.[78] Richel and Jacquet[79] reviewed the microwave pretreatment of lignocellulosic biomass, and they concluded that microwave technique favors the acceleration in reaction rate and improves the selectivities and yields.

2.6 ULTRASOUND PRETREATMENT

The pretreatment of lignocellulosic biomass using ultrasound technique is employed to extract lignin, hemicellulose, and cellulose from biomass materials, but research was not addressed extensively to study the susceptibility of lignocellulosic biomass to hydrolysis.[80] In a typical ultrasound technique, ultrasound is produced by a transducer fabricated from a piezoelectric material. The piezoelectric material generates characteristic mechanical vibration of ultrasonic frequency in response to an alternating current. It was revealed that saccharification of cellulose is augmented effectively using ultrasound pretreated lignocellulosic biomass.[81] Ultrasound influences cavitation caused by the introduction of ultrasound field into the enzyme processing solution and increases the movement of enzyme towards the surface of the substrate. Moreover, mechanical impacts, created by the breakdown of cavitation bubbles, give a consequential advantage of exploring the surface of solid substrates towards the activity of enzymes. It was published that better effects of cavitation can be brought about at 50°C, which is the reliable operating temperature for numerous enzymes.[81] Karimia et al.[82] reviewed the ultrasound irradiation pretreatment for the synthesis of ethanol from cellulosic biomass and starch-based feedstock, and they summarized that this technique facilitates the efficiency of hydrolysis and subsequently increases the sugar yield. It also highlights characteristics such as accessibility, crystallinity, degree of polymerization, morphological structure, swelling power, particle size, and viscosity as affected by ultrasonic treatment. Luo et al.[83] reported that sonication facilitates hydrolysis, esterification, and transesterification in biodiesel synthesis and leads to a decrease in reaction time by 50–80%, lower reaction temperature, fewer amounts of solvent than comparable unsonicated reaction systems.

2.7 PULSED ELECTRIC FIELD PRETREATMENT

In a typical pulsed electric field (PEF) pretreatment method, a short burst of high voltage is applied to a sample kept between two electrodes. Many useful applications of high electric fields for reversible or irreversible permeabilization of different biological processes are studied in the fields of medicine and bioscience.[84–89] The structure of plant tissues can be modified effectively using PEF pretreatment. A critical electric potential is generated across the cell membrane using high intensity, external electric field for nanoseconds to microseconds, and it favors quick electrical breakdown and local structural modification of the cell membrane, the cell wall, and therefore the plant tissue. An impressive increase in mass permeability and sometimes mechanical rupture of the plant tissue are favored using the electric field. The permeabilization of plant membranes to enhance mass transfer of metabolites is of interest to the food industry,[90] and inactivation of enzymes can be carried out using PEF pretreatment.[91] PEF technique is also used on vegetable tissue to increase the mass transfer processes such as diffusion of soluble substances,[92] juice extraction,[93,94] and dehydaration.[95] The electric field strength, the numbers of pulses, and the treatment duration influence the plant processing. In a typical PEF treatment of lignocellulosic biomass for biofuel production, the high field strength in the range of 5–20 kV/cm is applied for a microsecond to expose the cellulose in the plant fibers and to make permanent pores in the cell membrane, which facilitate the accessibility of acids or enzymes for breaking down the cellulose into its constituent sugars. The advantages of PEF pretreatment are as follows: it can be operated at ambient conditions, energy requirement is low as pulse times are very short (~100 μs), and there are no moving parts in the system.[96] It was reported that higher rate of hydrolysis was achieved using PEF pretreated switch grass at 2000 pulses of 8 kV/cm with a pulse width of 100 μs and a frequency of 3 Hz than untreated samples.[97] Zbinden et al.[98] investigated PEF pretreatment to improve lipid extraction from Ankistrodesmus falcatus wet biomass using the green solvent, ethyl acetate, and it was found that the extraction efficiency about 83–88% was lower compared to chloroform without using pulse electric field technique in 2-h incubation period. Yu et al.[99] studied the influence of PEF pretreatment on the valorization of extractives (proteins and polyphenols) from rapeseed green biomass (stems) by pressing. The influence of pressure, electric field strength, and pulse number on the juice expression yield, total polyphenols, and total proteins content in the expressed juices was investigated, and it was reported that PEF increases the juice expressed yield from 34% to 81% with significance increases in total polyphenols content, total proteins content, and in consolidation coefficient.

3

PHYSICOCHEMICAL PRETREATMENTS

3.1 STEAM EXPLOSION

Steam explosion is one of the most common and simple pretreatment methods for lignocellulosic biomass and is one of only a very limited number of economical pretreatment methods that have been upgraded to pilot scale affirmation and commercialized application.[21] In this method, steam explosion, aqueous separation, and hot water systems are involved. Stake Technology Ltd., Canada, developed the steam explosion technique, which includes the extrusion of the lignocellulosic biomass at a high temperature and pressure.[32] For a typical operation, biomass is treated with high-pressure saturated steam, and then the pressure is suddenly reduced, which causes the biomass to undergo an explosive decompression. Generally, the process is carried out at a temperature between 160 and 260°C and 0.70–4.8 MPa pressure for a few seconds to a few minutes before the biomass material is exposed to atmospheric pressure.[17] For increasing hemicellulose hydrolysis, the lignocellulosic biomass and steam mixture is held for a certain duration, and the process is completed with explosive decompression. The process favors hemicellulose degeneration and lignin alteration due to elevated temperature, thus enhancing the potential of cellulose hydrolysis. It has been reported that hemicellulose can be hydrolyzed by acetic acid and other acids generated during the steam explosion process. It was revealed that 90% efficiency of enzymatic hydrolysis was achieved in 24 h using steam explosion-pretreated poplar chips, whereas only 15% efficiency was observed in the case of untreated chips.

It has been shown that enzyme accessibility to the cellulose microfibrils can be augmented by exposing the cellulose surface through the removal of hemicellulose from microfibrils.[100] Though removal and redistribution of hemicellulose and lignin enhance the volume of the pretreated sample, complete removal of lignin is

DOI: 10.1201/9781003203414-3

difficult due to the redistribution of lignin on the fiber surface through melting, depolymerization, or polymerization reactions during the pretreatment process.[101] Accessible active surface area can be increased by creating a turbulent flow and rapid flashing of the biomass material to atmospheric pressure.[102] It has been reported by many researchers that water behaves as an acid at high temperatures during the pretreatment process.[22,103,104] Complete removal of hemicellulose can be performed by adding H_2SO_4 or SO_2 or CO_2 in the steam explosion process and the addition of supplementary chemicals significantly reduces the processing time and temperature boosts hydrolysis rate and resists the synthesis of inhibitory compounds.[105,106] It has been revealed that the addition of an acidic catalyst in steam explosion pretreatment of softwoods is essential for increasing the substrate accessible to enzymes.[14,102,106] Steam allows heating cellulose to the desired temperature rapidly while avoiding the immoderate dilution of the resulting sugars, and sudden pressure release decreases the temperature and quenches the reaction when pretreatment completes. The pretreatment time, temperature, particle size, and moisture content greatly affect the performance of the steam explosion process.[102,107] High hemicellulose degradation and hydrolysis rates were observed upon carrying out the steam explosion process either at a high temperature (ca. 270°C) and short treatment duration (ca.1 min) or at low temperature (ca. 190°C) and long treatment duration (10 min).[102] Low energy requirement in comparison with mechanical comminution is the main advantage of the steam explosion pretreatment, and no recycling or environmental costs are required. It has been reported that standard mechanical methods need 70% extra energy than steam explosion techniques to attain the same particle size reduction.[108] Addition of a catalyst to the steam explosion process significantly improves the performance. Upgrading the technique to a commercial scale is currently under study.[108] Steam explosion pretreatment is regarded as a highly economical process for agricultural residues and hardwood, but not so much for softwood.[17] Conversion of bamboo into methane through an improved fermentation process was studied using the steam explosion technique.[109] Sludge from a sewage treatment plant has been used as a microbial seed for methane fermentation. Methane production has been shown to increase by steam explosion even when not synthesized from raw bamboo. A correlation is observed between the methane synthesized and the high molecular weight of lignin. Ballesteros et al. studied the influence of the particle size of herbaceous lignocellulosic biomass on steam explosion pretreatment.[110] It was reported that a larger particle size (8–12 mm) shows a higher cellulose yield than the small particle size particle. 25% higher digestibility was observed using steam explosion-pretreated wheat, barley, and oat straws.[111] Cara et al.[112] investigated the synthesis of ethanol from olive tree pruning using the steam explosion pretreatment method with and without pre-soaking the biomass in water or sulfuric acid at varying temperatures. They noticed that the water presoaked sample treated with steam explosion technique at 240°C shows a higher ethanol yield (7.2 g of ethanol/100 g of raw material) on enzymatic hydrolysis followed by simultaneous saccharification and fermentation. The disadvantages of the steam explosion technique include incomplete depolymerization of the lignin, production of inhibitory substances that affect the performance of microorganisms used in the downstream processes, and demolition of a part of the xylan fraction.[113]

Washing of steam explosion-pretreated biomass and water-soluble hemicellulose with water is essential to remove the inhibitory substances that affect the performance of enzymatic hydrolysis and the rate of fermentation and microbial growth.[21] But washing with water reduces the overall saccharification yield through the removal of soluble sugars synthesized from hydrolysis of hemicellulose. Sometimes, liquid water is used for pretreatment while applying high pressure to keep the water in the liquid state at high temperatures.[114,115] This kind of pretreatment is called hydrothermolysis,[114] uncatalyzed solvolyses,[115] aqueous or steam/aqueous fractionation,[116] and aquasolv.[117] The pretreatment duration is maintained at 15–20 min at temperature 200–230°C.[118] Gonzalez et al.[119] carried out the enzymatic hydrolysis using liquid hot water and steam explosion-pretreated olive tree pruning biomass, and they reported that 75% and 45.5% sugars were synthesized from steam explosion pretreated sample and the hot water pretreated sample, respectively.

The influence of different parameters – such as temperature, pressure, pretreatment duration, and solid concentration on liquid hot water – was investigated and it was revealed that the influence of pretreatment duration in hemicellulose-derived sugar recovery in the prehydrolyzate relies on temperature, and enzyme hydrolysis yield increases as both temperature and time are increased.[120] Tabata et al.[121] conducted the steam-explosion of rice husk followed by enzymatic fermentation with Escherichia coli KO11, and it was found that no significant fermentation was noticed using the reducing sugar solution obtained by enzymatic saccharification of steam-exploded rice husk, but significance fermentation was achieved using hot water washed steam-exploded rice husk to ethanol in M9 fermentation medium (without glucose) because hot water favors removing inhibitory substances. Auxenfans et al.[122] studied the effect of steam explosion on the structural and chemical features of three different model biomasses (Miscanthus x giganteus, poplar, and wheat straw) and they reported that significant improvement of the cellulose conversion was achieved using combined steam explosion pretreatment with dilute sulfuric acid-impregnated samples. Wood et al.[123] investigated the influence of steam explosion on the enzymatic saccharification and simultaneous saccharification and fermentation of rice straw and husk at temperature 180–210°C for 10 min into hot water. Monschein and Nidetzky[124] studied continuous steam explosion on wheat straw and the influence of pretreatment intensity at varying acid loadings on hydrolysis of wheat straw using *Trichoderma reesei* cellulases. Ferro et al.[125] reported that 75% glucose yield was found using alkaline extraction after steam explosion pretreatment of rockrose and saccharification of untreated rockrose provided only 0.9% of sugars yield. It was also published that bioethanol production using simultaneous saccharification and fermentation (SSF) mode was a faster and more effective process than separate enzymatic hydrolysis and fermentation (SHF), and an ethanol concentration of 16.1 g/L with fermentation efficiency of 69.8% was observed. Bauer et al.[126] carried out the steam explosion pretreatment on hay and different pretreatment conditions were investigated. The highest glucose yield was obtained from the pretreated hay at 220°C for 15 min whereas higher xylose yields were found at 175°C for 10 min pretreatment time, and around 16% higher methane yield was achieved using steam-exploded hay than that of untreated hay. Iroba et al.[127] studied the steam explosion on barley straw grind in order to facilitate the decrystallization and depolymerization of the matrix, so as to have access to the hemicellulose and cellulose at a temperature in the range of 140–180°C, pressure in the range of 500–1100 kPa, the residence time of 5–10 min and mass fraction of water 8–50%. It was noticed that steam explosion favored the breakdown of biomass matrix with an increase in acid-soluble lignin but a considerable thermal degradation of cellulose and hemicellulose was observed.

3.2 AMMONIA FIBER EXPLOSION (AFEX)

A distinct physiochemical technique referred to as ammonia fiber explosion (AFEX) provides prospective as a pretreatment for lignocellulosic biomass. In the AFEX pretreatment method, lignocellulosic biomass is exposed to liquid ammonia at elevated temperature and pressure for a period of time, and then the pressure is suddenly lowered. The AFEX technique is quite similar to the steam explosion technique and it can remarkably increase the fermentation rate of different herbaceous grasses and crops.[128] Several types of lignocellulosic biomasses, such as alfalfa, wheat chaff, and wheat straw, can be treated using the AFEX technique.[129] A small amount of solid substance is solubilized while AFEX pretreatment is carried out and there is no hemicellulose or lignin removed. Oligomeric sugars are produced from hemicellulose and they are deacetylated, which may be the rationale behind hemicellulose not being soluble.[130] Higher digestibility is observed by increasing the water holding capacity through the structural change of the lignocellulosic biomass treated with the AFEX process.[13] It was reported that around 90% hydrolysis of cellulose and hemicellulose is synthesized using AFEX-pretreated Bermuda grass and bagasse.[131] On the other hand, it was published that the AFEX pretreatment process is not more effective for lignocellulosic biomass containing a higher lignin component, such as aspen chips, newspaper, woods, and nutshells, as for example, around 40% and 50% hydrolysis yields are observed using AFEX-treated newspaper and aspen chips, respectively.[21]

In the ammonia recycle percolation (ARP) method, aqueous ammonia (10–15 wt%) is charged into the lignocellulosic biomass with a fluid velocity of 1 cm/min and at a temperature range of 150–170°C for a pretreatment time of 14 min and after that, the ammonia is recovered and recycled.[13,17] In this process, aqueous ammonia interacts mainly with lignin and favors depolymerization of lignin polymer, and helps break the carbohydrate linkages in the lignin. No inhibitors for the downstream biological processes are synthesized as intermediate products and hence washing of the pretreated sample with water is not required in the ARP technique.[129] Teymouro et al.[132] studied the optimum process parameters, such as temperature, pretreatment duration, ammonia loading, and moisture content for AFEX pretreated corn stover. They published that ca. 98% theoretical glucose yield is found from enzymatic hydrolysis using AFEX-pretreated corn stover pretreated for 5 min at 90°C with ammonia to corn stover ratio of 1:1 and 60% (dry weight basis) moisture content of corn stover. They also indicated that temperature plays a significant role during AFEX treatment of lignocellulosic of biomass, as it regulates the system pressure favoring the volume of ammonia vaporized throughout the explosive flash. Higher rupture of the biomass fiber structure takes place at a higher ammonia vapor flush. Finally, the ethanol yield using the enzymatic hydrolysis of the AFEX pretreated corn stover enhanced about 2.2 folds than that of the untreated corn stover. Alizadeh et al.[128] carried out the study using AFEX-pretreated switchgrass at temperature ca 100°C, a pretreatment duration of 5 min, an ammonia to switchgrass ratio 1:1 (weight basis) and a moisture content of 80% (dry weight basis) and 93% glucan conversion was observed from enzymatic hydrolysis of AFEX pretreated switchgrass sample whereas 16% glucan conversion was found using untreated samples and the ethanol yield was 0.2 g ethanol/g of dry biomass, which was about 2.5 folds higher. Murnen et al.[133] carried out a study involving AFEX-pretreated water-soaked miscanthus at varying temperatures, ammonia to biomass loading, and pretreatment duration and it was revealed that about highest 96% glucan and 81% xylan conversions were found from enzymatic hydrolysis of the treated miscanthus at 160°C, an ammonia to biomass ratio 2:1(w/w), 2.3 g of water/g of biomass and 5 min pretreatment duration. It was reported that delignification study was carried out using aqueous ammonium hydroxide (30%)-soaked switch grass at varying ammonia to biomass ratios and pretreatment duration, and highest delignification, ca. 47%, was noticed using sample soaked in aqueous ammonia for 10 days with a loading of 10 mL ammonia/g switch grass.[134] Around 50% deglinification was achieved using corn stover soaked in aqueous ammonia for 4 days duration with a loading of 12 mL ammonia/g corn stover.[135] Kim et al.[136] published that 85% delignification was achieved using corn stover pretreated with ARP process. However, a higher hydrolysis rate, very near to theoretical yield, at low enzyme loadings (<5 FPU per gram of biomass or 20FPU/g cellulose) was reported using AFEX pretreated lignocellulosic biomass.[137] For making the AFEX process a more economical technique, efficient ammonia recovery is an essential step of the process and a viable approach is to recover the ammonia by evaporation just after the pretreatment process is completed.[138] Kamm et al.[139] studied the modified AFEX process for wheat straw to synthesize fermentable sugars. Aqueous ammonia (25% w/v) was used instead of liquid ammonia for investigating its effect on the sugar yield. It was observed that protein extraction just after the pretreatment can significantly improve the result found for the enzymatic hydrolysis and this modified AFEX process outlines simpler and less expensive parameters of the AFEX process usually described in literature. Ong et al.[140] carried out the AFEX process for corn stover and switch grass for investigating the impact of inter annual climate variability on biofuel production and hence corn stover and switch grass were collected during three years with significantly various precipitation profiles, representing a major drought year (2012) and two years with average precipitation for the whole season (2010 and 2013). All the samples were treated with AFEX method followed by enzymatically hydrolyzed and the hydrolysates were fermented separately with xylose-utilizing strains of *Saccharomyces cerevisiae* and *Zymomonas mobilis* and most corn stover and switch grass hydrolysates were readily fermented, and the growth of *S. cerevisiae* was completely inhibited in hydrolysate generated from drought-stressed switchgrass. This study revealed that variation in environmental conditions during the growth of bioenergy crops could have potential detrimental influence on fermentation organisms during biofuel production. Lee and Kuan[141] reviewed the species, cultivation, and lignocellulose composition of Miscanthus, pretreatment and enzyme saccharification of Miscanthus biomass for ethanol fermentation. They reported that a number of pretreatment techniques are used to increase the digestibility of Miscanthus biomass for enzymatic saccharification. A glucose yield of around 90% can be found using AFEX-pretreated miscanthus whereas a significant release of glucose yield can be possible using hot water or alkaline-pretreated sample and separation of valueable side products can decrease the overall production cost of bioethanol. Zhao et al.[142] studied the AFEX process of corn stover both soaking and without soaking in hydrogen peroxide for converting corn stover to fermentable sugars. It was found that higher delignification and enzymatic hydrolysis yields were found using hydrogen peroxide-soaked AFEX-pretreated corn stover that that of AFEX-pretreated corn stover only. Around 88% and 90% glucan and xylan were found from hydrogen peroxide soaked AFEX-pretreated corn stover at 0.7 (w/w) water loading, 1.0(w/w) ammonia loading, 0.5 (w/w) 30 wt.% hydrogen peroxide loading, and a temperature of 130°C for 10 min. The results from their study revealed that hydrogen peroxide followed by AFEX pretreatment is a potential pretreatment to improve the enzymatic saccharification of corn stover for bioethanol production.

3.3 CARBON DIOXIDE EXPLOSION

The rate and the extent of cellulose hydrolysis can be augmented by influencing the reactivity of cellulose though the pretreatment of lignocellulosic biomass with the carbon dioxide explosion technique. The carbon dioxide explosion technique is very similar to the steam and ammonia explosion techniques, and it was postulated that carbon dioxide produces carbonic acid and enhances the hydrolysis rate. The aim of utilizing the carbon dioxide explosion technique was to develop a better pretreatment for lignocellulosic biomass materials at a lower temperature than the steam explosion technique and a lower cost than the expensive ammonia explosion technique. It is postulated that carbon dioxide produces carbonic acid with water and this acid favors the hydrolysis rate. Carbon dioxide molecules are easily diffused into the pores accessible to water and ammonia molecules as their sizes are comparable to water and ammonia and decomposition of monosaccharides with acid can be resisted at low temperature. The higher active surface area of the substrate for hydrolysis is enhanced through the rupture of the cellulosic crystal structure under the explosive release of the carbon dioxide pressure. It was reported that 75% of the theoretical glucose was achieved after 24 h enzymatic hydrolysis of carbon dioxide explosion pretreated alfalfa with a loading of 4 kg of CO_2 per kg of fibre at a pressure of 5.62 MPa.[143] On the contrary, it was reported in a study carried out using carbon dioxide explosion, steam explosion, and ammonia explosion pretreated recycled paper, repulping waste of recycled paper and sugarcane bagasse.[144]

Narayanaswamy et al.[145] studied the super critical CO_2 pretreatment of corn stover and switch grass at different temperatures and pressures and the pretreated biomass was hydrolyzed using cellulase combined with b-glucosidase. It was found that the highest glucose yield, 30%, was achieved using supercritical CO_2-pretreated corn stover at a temperature of 150°C and a pressure of 3500 psi for 60 min than that obtained from untreated samples (12% glucose yield). Srinivasan and Ju[146] investigated the supercritical CO_2 pretreatment of guayule for finding out the optimum temperature, pressure, moisture, and treatment duration using a central composite design (CCD). Enzymatic hydrolysis was carried out using a pretreated sample at 30°C and an enzyme solution with 5% (w/v) solid loading for 72 h. The yields of glucose and pentose were measured and used as CCD response variables and statistical analysis of results recommended that 56% glucose and 61% pentose yields were achieved at 175°C temperature, 3800 psi pressure, 60% moisture, and 30 min. Statistical analysis of results led to the following recommended condition: 175°C, 26.2 MPa (3800 psi), 60% moisture, and 30 min. Gu et al.[147] studied the super critical CO_2 and ionic liquid pretreatments of different biomass materials such as corn stover, switch grass, sugarcane bagasse, soft and hardwood. Capolupo and Faraco[148] reviewed the most commonly used "green" pretreatment processes including, supercritical CO_2 pretreatment for bioconversion of lignocellulosic biomasses. Carneiro et al.[149] reported a literature review of biomass pretreatments that made use of carbon dioxide as a pretreatment agent. It was shown that CO_2 is used as supercritical carbon dioxide explosion and sub/supercritical water hydrolysis with carbon dioxide.

3.4 WET OXIDATION

The wet oxidation process is a technique in which water and oxygen are used at high temperatures for fractionation of lignocellulosic biomass. Sometimes air is used as an oxygen source. Wet oxidation pretreatment is carried out with a very short residence time and at a low temperature.[150] A typical wet oxidation is carried out at temperatures 170 to 200°C, pressures 10 to 12 bar oxygen and for residence time 10–15 min.[151] The wet oxidation process can be carried out in an exothermic manner at temperatures above 170°C for decreasing the total energy required. The main reactions associated with the wet oxidation process are the synthesis of acids from hydrolytic processes and oxidative reactions. It is an efficient technique to increase the digestibility of cellulose by solubilizing the hemicelluloses and lignin and it is used for ethanol synthesis followed by simultaneous saccharification and fermentation (SSF).[152] In this technique, phenolic compounds are not synthesized as end products as they are further converted into carboxylic acids. It was reported that 65% glucose yield was observed using enzymatic hydrolysis of wet-oxidation-pretreated wheat straw[153] and higher yields were found using wet-oxidation-pretreated corn stover[154] and spruce. In spite of production of a small amount of inhibiting substances during pretreatment, higher delignification is achieved using the wet oxidation method and the cost of oxygen is the major disadvantage for recommendation for using the wet air oxidation technique. Ayeni et al.[154] investigated the alkaline wet air oxidation and alkaline peroxide-assisted wet air oxidation of sawdust at various temperature, pressure, reaction time, and 518 g/kg cellulose content. 580 g/kg hemicellulose content and 171 g/kg lignin removal were achieved using wet air oxidation of the samples at 170°C temperature, 1.0 MPa pressure, 10 min pretreatment duration. Fang et al.[155] carried out the wet oxidation pretreatment of poplar residues for bioethanol production and optimizing the wet oxidation pretreatment parameters for using poplar residues from the paper mill industry and optimum conditions for treatment are as follows: initial pH:10, temperature: 195°C, oxygen pressure: 1.2 MPa and treatment time: 15 min. Srinivas et al.[156] investigated the wet explosion followed by wet oxidation of forestry residues of Douglas Fir (FS-10) in alkaline conditions and at 10% solids, 11.7 wt.% alkali and 15 min reaction time upon enzymatic hydrolysis yields of glucose 12.9%, vanillin 0.4wt% at 230°C; formic acid 1.6wt.% at 250°C; acetic acid 10.7 wt%, hydroxybenzaldehyde 0.2 wt%, syringaldehyde 0.13 wt% at 280 °C; and lactic acid 12.4 wt% at 300°C were observed. Recently, Katsimpouras et al.[157] carried out the wet oxidation of beech wood residual biomass employing hot acetone, water, and oxygen and it was found to be an attractive pretreatment technique for delignification. It created fewer degradation products.

4

CHEMICAL PRETREATMENTS

4.1 OZONOLYSIS

Ozonolysis is an effective pretreatment method for lignocellulosic biomass, and ozone is a strong oxidizing agent that exhibits high delignification efficiency.[158] This pretreatment method does not synthesize any toxic residues and enhances the in vitro digestibility of the treated biomass. The ozonolysis technique is applied to degrade hemicellulose and lignin in many lignocellulosic biomasses, such as wheat straw,[159] bagasse, green hay, peanut, pine,[160] cotton straw,[161] and popular sawdust.[162] It has been found that 60% delignification can be achieved using ozone-pretreated wheat straw, and the enzymatic hydrolysis rate is hiked 5 folds, whereas 29% delignification is observed using ozone-pretreated poplar sawdust and enzymatic hydrolysis yield is increased by about 57%.[162] The advantage of ozonolysis is that the process can be executed at normal temperature and pressure and environmental pollution due to the release of ozone can be minimized by decomposing ozone over a catalytic bed.[163] The ozonolysis technique is a little expensive, as it requires a large volume of ozone. Several research groups have reported synthesizing oxalic and formic acids along with glycolic, glycoxylic, glyceric, malonic, p-hydroxybenzoic, succinic, fumaric, and propanoic acids using ozone-pretreated poplar sawdust containing 45% moisture.[164,165] P-hydroxybenzoic, azelaic, malonic, vanillic, caproic, levulinic, acids, and aldehydes, such as vanillin, hydroquinone, and p-hydroxybenzaldehyde have been synthesized using ozone-pretreated herbaceous species containing 50% moisture.[166] Panneerselvam et al.[167] carried out the ozonolysis of Miscanthus giganteus, Miscanthus sinensis 'Gracillimus', Saccharum arundinaceum, and Saccharum ravennae, commonly referred to as 'energy grasses' at three ozone concentrations of 40, 50, and 58 mg/L using two ozone flow modes, unidirectional flow, and reverse flow. The process parameters for each sample were optimized on the basis of the lignin content and glucan recovery, and around 60% delignification was achieved effectively without any cellulose degradation. Enzymatic

DOI: 10.1201/9781003203414-4

hydrolysis of the pretreated sample with Cellic CTec-2 at 0.06 g/g raw biomass shows glucan conversion lower than that obtained from untreated samples due to the strong enzyme inhibition by the lignin degradation products generated during ozonolysis pretreatment. Souza-Correa et al.[168] investigated sugarcane bagasse pretreatment with ozone via atmospheric O_2 pressure plasma, and 80% delignification was achieved within 6 h of processing time without affecting the cellulose by ozonolysis, and with the removal of some hemicellulose. It was reported that moisture content favored the delignification of bagasse, where 50% moisture content was recognized as optimum and 65% of the cellulose was transformed into glucose. It was found that ozone molecules break the strong carbon-carbon bonds of aromatic rings slower than the weak carbon bonds of aliphatic chains. Bellido et al.[169] studied the acetone-butanol-ethanol (ABE) synthesis using ozonation of wheat straw, and it was observed that ca. 65% glucose and ca. 40% xylose yields were achieved using enzymatic hydrolysis of ozonated wheat straw and slightly lower around 0.32 g/g ABE yield and 79.65. g. ABE/kg wheat straw was achieved upon fermentation by *Clostridium Beijerinckii* of ozonated wheat straw hydrolysates compared to that obtained from steam-exploded wheat straw hydrolysates. The influence of the predominant inhibitory substances indicated in hydrolysates (oxalic acid, acetic acid, 5-hydroxymethyl furfural, and furfural) was looked into through ABE fermentation in model media. A simulation of the ozone pretreatment of wheat straw was proposed to study the change in lignin concentration with different ozonolysis times.[170] Li et al.[171] carried out the ozonolysis of maize stover to improve the enzymatic digestibility and study the influence of sample particle size and moisture content on the ozonolysis process. It was observed that the highest 75% delignification was achieved at the mesh of -300, and the moisture of 60%, and that the smaller particle size favors the process and free and bound water ratio plays a key role for ozonolysis. The glucose yield through enzymatic hydrolysis ranged from 18.5% to 80% and water wash potentially reduced the xylose yield (up to 42% decreases) but the influence on glucose yield was low (<10% increases). Moreover, xylan solubilization during ozonolysis and a simulation of the ozone pretreatment of wheat straw were noticed. Recently, Travaini et al.[172] reported that 84% glucose and 67% xylose yields were found from ozonated and water-washed sugarcane bagasse and concentration of ozone showed the highest influence on delignification and sugar release after washing than the other studied process parameters, such as moisture content, particle size, ozone, and oxygen flow. It was also reported that the moisture content of the sample potentially influences the formation of inhibitory substances during the process. Finally, 88% ethanol yield was achieved using Pichia stipitis, whereas Clostridium acetobutylicum yielded 0.072 g butanol/g of sugar.

4.2 ACID PRETREATMENT

Acid pretreatment is a potential technique for the depolymerization of low-lignin lignocellulosic materials, and it can favor high yields of hemicellulosic sugars and can enhance the enzymatic yields of glucose for a biomass-to-biofuels process. The anaerobic digestibility can be increased using acid-pretreated lignocellulosic biomass. Solubilization of the hemicellulose is carried out using acid pretreatment and it makes the cellulose more accessible to enzymes. Usually, mineral acids such as HCL and H_2SO_4 are used in the acid pretreatment technique. Sometimes, lignocellulosic biomass is pretreated with dilute acids, such as dilute sulfuric acid, and this pretreatment technique is a single-step process. A two-step process was used in National Renewable Energy Laboratory (NREL) in Golden, Colorado, for the reduction of enzyme loading. It was reported that the efficiency of delignification and the sugar yield upon enzymatic hydrolysis of rice straw pretreated with dilute acid was not satisfactory.[173,174] Concentrated acids, such as HCl and H_2SO_4, are strong agents for cellulose hydrolysis, but they are highly corrosive, toxic, and hazardous. So, it requires reactors that are resistant to corrosion, which makes the process more expensive. It has been reported that for making the acid pretreatment process more economical, recovery of acids after hydrolysis is essential.[17,175] Furfural was synthesized commercially using dilute H_2SO_4 acid-pretreated lignocellulosic biomass and acid concentrations usually less than 4 wt.% economically feasible.[176,177] In this process, dilute H_2SO_4 is added with lignocellulosic biomass to hydrolyze hemicellulose to xylose and other sugars and then extend to crack xylose down to synthesize furfural. A higher rate of cellulose hydrolysis is reported using the dilute H_2SO_4 pretreatment technique[178] and a high temperature favors the cellulose hydrolysis rate.[21] High xylan-to-xylose conversion yields were observed using advance dilute acid hydrolysis at lower harsh conditions, and this is crucial to attaining favorable overall process economics because xylan exists about one-third of the total carbohydrate in many lignocellulosic biomasses.[179]

Two kinds of dilute acid pretreatment techniques are used: continuous flow mode for lower solid loadings at a temperature higher than 160°C and batch mode for higher solid loadings at a temperature lower than 160°C.[178,180,181] The dilute sulfuric acid pretreatment process is used widely, however, dilute nitric acid,[182] dilute hydrochloric acid[183,184] and dilute phosphoric acid[183] pretreatments are also used for corn (husks, cobs, stover) and hardwood bark from aspen, poplar, and sweet gum. It is reported that a higher crytallinity index was noticed using dilute sulfuric acid pretreated mixed hardwood, although crystallinity index is not influenced by the pretreatment temperature.[185] The influence of porosity on enzymatic digestibility of the cellulose was studied using dilute sulfuric acid pretreated corn stover in a pilot-scale vertical reactor at temperatures ranging from 180 to 200°C, acid loadings of 0.03 to 0.06 g of acid per g of dry biomass, and solid loadings between 25% to 35% (w/w) for a fixed pretreatment time of 1 min.[186] It was revealed that pretreated corn stover exhibits a higher pore volume than untreated samples. Finally, it was inferred that porosity does not have any significant influence on higher digestibility but it may influence the lignocellulosic biomasses with low digestibility. It was also reported that 70% glucose yield was found using 2% dilute sulfuric acid pretreated corn stover at 120°C for 43 min pretreatment duration.[187] Cara et al.[188] studied the dilute sulfuric acid pretreatment of olive tree biomass at varied acid concentrations of 0.2%, 0.6%, 1%, and 1.4% (w/w) and temperatures ranging from 170 to 210°C. They reported that a maximum of 83% of hemicellulosic sugars was achieved using 1% (w/w) sulfuric acid concentration at 170°C and a maximum of 76.5% enzymatic hydrolysis yield was observed using 1.4% (w/w) sulfuric acid at 210°C. Yat and co-workers[1] carried out dilute acid pretreatment of four timbers (aspen, basswood, red maple, and balsam) and switchgrass at temperatures of 160 to 190°C, dilute sulfuric acid concentrations of 0.25 to 1% (w/v), particle sizes of 28 to 10/20 mesh in a glass-lined 1 L batch reactor. They revealed that temperature and acid concentration strongly influence the rates of formation of xylose from hemicellulose and the formation of furfural from xylose and maximum yields ranging from 70% (balsam) to 94% (switchgrass) for xylose, from 10.6% to 13.6% for glucose. Though dilute acid pretreatment favors cellulose hydrolysis effectively, it costs higher than physicochemical pretreatments such as a steam explosion or AFEX. Also, the downstream enzymatic hydrolysis or fermentation processes require pH neutralization for efficient performance. The synthesis of spherical droplets on the surface of the residual corn stover pretreated under neutral pH and acidic pH and temperature above 130°C was noticed and it was investigated that droplets contain lignin and possible lignin carbohydrate complexes.[189] This kind of droplets decrease the enzymatic saccharification of the substrate. It is also revealed that some toxic substances generated during dilute acid pretreatment affect the performance of the fermentation.[13] Moreover, acid pretreatment uses high pressure, and hence it requires expensive materials for fabrication. Other than that, some more steps such as neutralization and conditioning of hydrolysate before biological processes are the essential, and the slow rate of cellulose digestion by enzymes and nonproductive binding of enzymes to lignin are the major drawbacks of the methods.[190] Kumar et al.[191] carried out the acid pretreatment of de-oiled jatropha waste and the highest xylose yields of 6.89g/L and 6.16g/L were achieved using HCl and H_2SO_4 pretreated jatropha waste, respectively, and a production of 3.1 L H_2/L reactor was achieved using combined acid and enzyme hydrolysis in the batch mode operation. Adaganti et al.[192] reported that 2.67 g/L and 1.72 g/L glucose were synthesized with H_2SO_4 and HCl acid hydrolysis, respectively, for hydrothermal exploited Calliandra calothyrsus biomass and maximum bioethanol yield was obtained at pH 4.5, temperature 30°C, and incubation period of 72 h. Swaminathan et al.[193] carried out acid pretreatment of sweet sorghum stalk using different dilute acids for effective delignification and hydrolysis and for optimizing the simple monomers yield at various acid concentrations. The pretreatment was performed with different acids such as HCl and H_2SO_4 at various concentrations of 0.5% (v/v) to 3% (v/v) to maximize the sugars yield. The maximum sugar yield was observed using 1% (v/v) HCl pretreated sweet sorghum sample hydrolysate. Scaninf Electron Microcope (SEM) analysis showed that the lengthier microfibrils structure of the biomass network was modified and more hole formation on the annular rings and on the cellulose wall, which were the internal frame of the biomass network, was observed. Muktham et al.[194] investigated the acid pretreatment of de-oiled pongamia pinnata seed cake at varying acids such as H_2SO_4, HCl, H_3PO_4, acid concentration of 2–6% w/w, and temperature of 80–100°C. It was revealed that acid concentration and temperature exhibited a positive effect on glucose release from the biomass using HCl acid as the best catalyst in comparison with H_2SO_4 and H_3PO_4 and the highest glucose yield of 173.4 g. kg seed residue was obtained at 100°C with 6% w/w HCl concentration and finally, 88.62 g ethanol/kg dry seed residue was achieved spending the energy required for this pretreatment was estimated to get an insight into the process energy demand (1080–1110 KJ/kg of seed cake). Recently, Mohapatra et al.[195] studied the acid and ultrasono-assisted acid pretreatments of two Pennisetum sp.: Denanath gras (DG) and Hydrid Napier grass (HNG) using various acids such as H_2SO_4, HCl, H_3PO_4, and H_2NO_3 at different temperatures, variable soaking times and acid concentrations applying Taguchi orthogonal array and the data generated were statistically validated using artificial neural networking. For effective acid pretreatment, HCl was indicated as the best acid for both the Pennisetum sp. and highest delignification yields 33% for DG and 33.8% for HNG were achieved at 1% and 1.5% of HCl acid concentrations, temperature 121°C, and 110°C and 130 min and 50 min soaking duration respectively. Increased delignification yields of 80.4% for DG and 82.1% for HNG were observed using ultrasonic-assisted HCl pretreatment with a power supply of 100W, the temperature of 80°C, and duty cycle of 70%.

4.3 ALKALI PRETREATMENT

Alkaline pretreatment is an extensively used chemical pretreatment technique. It is mostly a delignification technique but some amount of hemicellulose is also solubilized in this method. The process mechanism is postulated to be saponification of intermolecular ester bonds crosslinking xylan hemicellulose and other components such as lignin and alternative hemicellulose. This method decreases the synthesis of inhibitory substances such as acetyl and different uronic acids that decrease the accessibility of hemicellulose and cellulose to enzymes.[196] Alkali pretreatment technique uses the alkaline solutions such as NaOH or KOH to deconstruct the lignin and hemicellulose, and hence enhance the accessibility of enzyme to the cellulose. This pretreatment can be carried out at a low temperature and pressure and relatively long pretreatment duration and high concentration of base[9] and the influence of alkaline pretreatment depends on the lignin content of lignocellulosic biomass.[21,30] In comparison with acid or oxidative reagents; alkali pretreatment method is considered as the best efficient method for cleavage the ester bonds between lignin, hemicellulose and cellulose, and preventing the depolymerization of hemicellulose.[196] Alkaline technique resists sugar degradation in comparison with acid pretreatment methods and recovery and regeneration performances of caustic salts are higher. Sodium, calcium, potassium, and ammonium hydroxides are suitable reagents of which sodium hydroxide uses extensively.[197–200] But, it was reported that efficient pretreatment was carried out using calcium hydroxide (slake lime) and it was very cost effective per kg of hydroxide.[201] In a typical lime pretreatment, the particle size of the biomass ranges 5–10 mm and lime and water slurry is sprayed onto the biomass then biomass material in a pile is kept for a period of hours to weeks. The correlation between enzymatic digestibility and structural aspects such as crystallinity, lignin, and acetyl content was studied and it is reported that high digestibility can be achieved using comprehensive delignification regardless of crystallinity and acetyl content; deacetylation and lignin deploymerization favor enzymatic hydrolysis; and hydrolysis rate is affected by crystallinity initially but it affects slightly on final sugar yields.[202] Lee and co-workers[203] published that enzyme adsorption and the efficacy of the adsorbed enzymes influence the rate of the hydrolysis rather than the diffusive mass transfer of enzyme. Elimination of nonproductive adsorption sites through delignification process augments enzyme effectiveness towards cellulose and hemicellulose. It was reported that the reduction of the steric hindrance and augmentation of carbohydrate digestibility can be done by removing acetyl groups from hemicellulose (mainly xylan) under alkaline pretreatment.[204] It is also inferred that acetyl group content affects the sugar yield. Several numbers of research groups studied the lime pretreatment of wheat straw (85°C for 3 h),[205] poplar wood (150°C for 6 h with 14 atm of oxygen),[206] switchgrass (100°C for 2 h),[207] and corn stover (100°C for 13 h).[208] Karr and co-workers[209] carried out the study using slake lime pretreated corn stover at varying temperature, lime loading, water loading, and pretreatment duration and it was noticed that enzymatic hydrolysis of pretreated sample is 9 folds higher than that of untreated sample at 120°C, a lime loading of 0.075 g of $Ca(OH)_2$/g of dry biomass, a water loading of 5 g of H_2O/g of dry biomass for 4 h pretreatment time. It was also reported that higher hydrolysis yield is observed using NaOH pretreated straws with comparatively low lignin contents of 10–18%.[210] Chosudu et al.[211] carried out the study with combination of irradiation and 2% NaOH alkaline pretreatment of corn stalk, cassava bark, and peanut husk and they reported that around 43%

and 20% glucose yields were found from alkaline and electron beam irradiation pretreated sample and untreated sample respectively whereas around 3.5% and 2.5% glucose yields were observed using combined pretreatment of cassava bark and peanut husk respectively. It is published that radio frequency assisted alkaline pretreatment of switch grass was carried out at temperature 90°C, 0.1–0.25g of NaOH/g of biomass loading with size of the biomass particle 0.25–5.0 mm.[212] The maximum yields for glucose and xylose were found 37.3 g/100 g of biomass and 21.9 g/100 g of biomass, respectively at temperature 90°C and 0.25g of NaOH/g of biomass loading. It was revealed that the microwave-assisted and conventional-heated alkaline pretreatment of switch grass was carried out at 0.053–0.3 g of alkali/g of biomass loading and higher sugar yield was achieved using microwave-assisted alkaline pretreatment of the sample than conventional heating and 90% sugar yield was found using microwave-assisted heating at an alkali loading of 0.1 g/g.[213] This study inferred that enzymatic digestibility of switch grass can be enhanced using microwave-assisted alkali pretreatment technique. It is further studied that 60 to 80% deleginfication for corn stover and 65 to 85% delignification for switch grass were observed using ammonia recycled percolation pretreatment for 1 h reaction time at 170°C and ammonia concentration of 2.5–20%.[214] Kataria et al.[215] carried out the NaOH pretreatment of Kans grass at various concentrations of NaOH of 0.5%, 1%, 1.5%, and 2%, different pretreatment time of 30, 60, 90 and 120 min and different temperatures of 100, 110 and 120°C. Above 50% delignification was observed at 120°C and subsequently sugar yield of 350 mg/g dry biomass was achieved from enzymatic saccharification using crude enzyme (obtained from Trichoderma reesei), which was 5 folds higher than enzymatic saccharification of acid pretreated biomass (69.08mg/g) as reported elsewere.[216] The fermentation of enzymatic hydrolysate using microbes showed 0.44–0.46gg-1 ethanol yield and no toxic compounds were found in comparison to acid pretreatment technique. Gao et al.[217] carried out the NaOH alkaline pretreatment of energy crop (switch grass) and an invasive (North America) plant species (phragmites) and sugar yields 365 and 385 g/kg raw biomass were obtained at concentration of 1% NaOH, temperature of 121°C for 30 min treatment duration using switch grass and phragmites respectively. Enzymatic hydrolysis was carried out in a 125 mL erlenmeyer flasks using Cellic CTec-2 cellulase (Novozyme) and an air-bath incubator (Infors Multitron, Infors

Switzerland) at 50°C and 150 rpm for 72 h. Acetone butanol ethanol (ABE) yields of 146 and 150 g/kg dry plant biomass were achieved from alkaline treated switch grass and phragmites respectively through fermentation process using *C. saccharobutylicum* DSM 13864. However, similar overall solvent yields were synthesized from both samples and the higest carbon loss occurred for switch grass during pretreatment but the highest loss in the case of phragmites occurred to enzymatic hydrolysis. Stoklosa et al.[218] studied the alkaline pretreatment of woodchips with 18% NaOH (w/w) and 4:1 liquor-to wood ratio and around 50% of the xylene was solubilized and underwent degradation, depolymerization, and substantial removal of the glucuronic acid substitutions on the xylan during the bulk delignification process. An interesting finding was that substantial xylan was found in the liquor without degradation. Cellulose hydrolysis yields of around 80 to 90% were achieved within 24 ton 48 h using alakali pretreated diverse hardwoods at modest enzyme loadings and initial rate of hydrolysis was increased with decrease in lignin content. Siddhu et al.[219] investigated the pretreatment of corn stover using alkali to modify its recalcitrance nature for biofuel synthesis. In their experiment, 1.5% KOH and its black liquor (spent liquor of KOH) were used to pretreat corn stover (CS) at temperature of 20°C to enhance the digestibility for anaerobic digestion (AD). It was found that no remarkable difference was observed in weighted average methane content on the basis of methane and biogas yields between black liquor and original KOH treated corn stover after anaerobic digestion. The black liquor treatment favored the overall methane yield by 52.4% in comparison with untreated corn stover (135.2 ml/g VS), whereas no remarkable difference between the overall methane yields of 1.5% KOH treated and black liquor treated corn stover was noticed. However, significantly 56% water and 57.4% KOH were saved using black liquor technique in comparison with the 1.5% KOH pretreatment. Therefore, recycling and reuse of KOH black liquor would be an effective method for lignocellulosic biomass pretreatment and future anaerobic digestions. Recently, Ling et al.[220] explored the crystallinity and crystalline structure of alkaline pretreated cellulose and subsequently enzymatic hydrolysis of the pretreated cellulose was performed for studying the effieciency of sugar conversion. Alkaline pretreatment of cellulose Iβ using less than 8 wt% NaOH favored higher cellulose crystallinity and deploymerirized hemicelluloses, that were superimposed to influence the enzymatic hydrolysis to glucose and changing crystallite sizes and lattice spacing confirmed the decrystallization of cellulose structure under pretreatment with 8 to 12 wt.% NaOH.

4.4 PRETREATMENT WITH OXIDIZING AGENT

In this pretreatment method, some additional highly oxidizing agents such as hydrogen peroxide, peracetic acid are used to suspended biomass in water. Delignification and removal of hemicellulose from lignocellulosic biomass materials can be enhanced using oxidative pretreatment method. Several reactions such as electrophilic substitution, side chains displacement, breaking of linkages between alkyl and aryl and oxidative breaking of aromatic nuclei are taken place during oxidative pretreatment.[221] Hydrogen peroxide pretreatment method favors the oxidative delignification to depolymerize and solubilize the lignin and loosen the cellulosic crystal matrix thus boosting enzyme accessibility to the substrate.[222] Wei and Cheng[223] carried out the hydrogen peroxide pretreatment of rice straw for studying the influence of pretreatment on the change of the structural matrix and the enzymatic hydrolysis. To explore the modification of the structural feature of the rice straw, various parameters such as lignin content, water holding capacity, weight loss, crystallinity of straw, and accessibility of Cadoxen were investigated and for finding out the influence of pretreatment on rice straw different parameters such as cellulose adsorptions, cellobiose accumulation in the initial stage of hydrolysis and rates and extents of enzymatic hydrolysis were studied. They reported that about 60% delignification, 40% weight loss, a 5 folds augmentation in the accessibility for Cadoxen, 1 fold augmentation in the water holding capacity and small decrease in crystallinity in comparison with that of the untreated rice straw at temperature 60°C and with 1%(w/w) H_2O_2 and NaOH solution for 5 h treatment duration. The initial rates and extents of hydrolysis, cellulose adsorption, and cellobiose accumulation in hydrolysis were increased as per the modification of the structural matrix of pretreated rice straw and enzymatic hydrolysis extent was enhanced bout 4 folds for 24 h hydrolysis time using alkaline H_2O_2 pretreated rice straw. It was reported that the degree of enzymatic solubilization relative to the amount of residual straw was ca. 42% using 20% per acetic acid alikaline pretreated rice straw and negligible rupture of the crystalline structure of cellulose in the straw was reported.[224,225] The pretreatment of can bagasse using hydrogen peroxide was investigated and it was published that lignin biodegradation was increased by the peroxidase enzyme in the presence of H_2O_2.[226] About 50% of the lignin and all hemicellulose were solubilized with 2% H_2O_2 at 30°C within 8 h pretreatment duration and finally 95% efficiency of glucose synthesis was observed using the subsequent saccharification by cellulose at 45°C for 24 h. Bjerre and co-workers[210] studied the wet oxidation and alkaline hydrolysis of wheat straw and it was revealed that 85% conversion yield of cellulose to glucose was found at 170°C for 1–10 min. They reported that the delignification of wheat straw was enhanced and no inhibitory substances such as furfural and hydroxyl methyl furfural were synthesized during wet oxidation combined with base addition and thus making the polysaccharides more susceptible to enzymatic hydrolysis. Banerjee et al.[227] investigated the alkaline hydrogen peroxide pretreatment of corn stover at 22°C temperature, 0.125g H_2O_2/g biomass and atmospheric pressure for 48 h with periodic pH adjustment in order to study the feasibility of scaling-up the process and integrating it with enzymatic hydrolysis followed by fermentation. About 75% glucose and 71% xylose were synthesized and the final ethanol titer was 13.7 g/L using *Saccharomyces Cerevisiae* for 120 h. It was also reported that xylose fermentation was enhanced significantly using activated carbon pretreatment of enzymatic hydrolysate prior to fermentation but little effect was observed on glucose fermentation, presumably due to the removal of soluble aromatic inhibitors. Alkaline hydrogen peroxide pretreatment shows significance advantages with respect to process simplicity, feedstock handling, capital costs, and compatibility with enzymatic deconstruction and fermentation in comparison with other leading pretreatments for lignocellulosic biomass. Loow et al.[228] carried out the pretreatment of oil palm fronds using divalent ($CuCl_2$) and trivalent ($FeCl_3$) inorganic salts and oxidizing agents such as hydrogen peroxide and sodium persulfate were used for improving the inorganic salt pretreatment.

4.5 ORGANOSOLVATION PRETREATMENT

The organosolvation method is a depolymerization technique, with varying concurrent hemicellulose solubilization. This is a favorable pretreatment technique for employment in lignocellulosic biomass pretreatment.[229] In a typical organosolvation process, internal lignin and hemicellulose bonds cleavage is taken place by interacting organic or aqueous organic solvent mixture such as methanol, ethanol, acetone, ethylene glycol, triethylene glycol and tetrahydrofurfuryl with inorganic acid catalyst such as HCl or H_2SO_4.[230,231] Sometimes organic acids such as oxalic, salicylic, and acetylsalicylic acids are used but they are toxic in nature.[232] The organosolvation process associated with the delignification of lignocellulosic biomass using organic solvents and, mostly dilute aqueous acid solutions followed by simultaneous prehydrolysis.[233] It was reported that addition of small amount acid enhanced the xylose yield and pulps with residual lignin of 6.4% to 27.4%(w/w) was synthesized using mixed softwoods and aqueous ethanol at temperature of 200°C and 400 psi pressure with ethanol to water ratio of 1:1 (w/w) for application of biorefining technology called the lignol process. Hydrolysis of pulps was performed efficiently without additional delignification and above 90% of the cellulose in low lignin pulps (<18.5% leftover lignin) was hydrolyzed to glucose within 48 h. It was reported that woody biomass was heated in water and ethanol liquor following the lignol process at a specific temperature and time and it was observed that various chemical hydrolysis reactions are occurred and different chemical constituents were synthesized.[234] It was reported that cellulose hydrolyzed partially into smaller fragments at temperature 195°C, an ethanol concentration 70% (w/w), a liquor to solids ratio 10:1(w/w), at pH = 3.8 and pretreatment time of 90 min. Under such conditions, hemicellulose hydrolyzed into oligosaccharides, monosaccharides, and acetic acid and production of acetic acid decreased the liquor pH and but favored the hydrolysis of other constituents and lignin hydrolyzed into lower molecular weight fragments that dissolved in the aqueous ethanol liquor. Pan and co-workers[235] studied the organosolvation treatment involving extraction with aqueous ethanol for the conversion of poplar chips to ethanol and they reported that cellulose rich solids fraction, an ethanol organosolvation lignin (EOL) fraction, a water soluble fraction containing hemicellulosic sugars, degraded lignin, and small amounts other constituents were synthesized during treatment process. In order to make the organosolvation treatment economic and cost effective, solvents are separated from the reactor, evaporated, condensed, and recycled, and quick removal of solvents from the system is essential as they act as inhibitory substances which decrease the growth of microorganisms, enzymatic hydrolysis, and fermentation.[235] Wildschut et al.[236] carried out the ethanol organosolv pretreatment of wheat straw focusing on various process parameters such as temperature, reaction duration, acid catalyst loading, solvent concentration, and particle size to study their affects on delignification, xylan hydrolysis, and enzymatic cellulose digestibility. The lignin yield about 84% was observed using 50% (w/w) aqueous ethanol at 210°C without using any catalyst and highest glucose yield of 86% was achieved from enzymatic hydrolysis of the pretreated sample using the commercial enzyme mixture Accellerase 1500 obtained from DuPont Industrial Biosciences (Leiden, NL). Interestingly, similar lignin and glucose yields were achieved using 30 Mm H_2SO_4 as catalyst at lower temperature of 190°C. By decreasing the temperature with an acid catalyst substantially favored the yield of the hemicellulose derivatives xylose and furfural. Bouxin et al.[237] studied the organosolv pretreatment of Sitka spruce sawdust using ethanol, water, and dilute sulfuric acid and subsequent saccharification yields ca. 86% was observed along with the conversion of large portion of the hemicellulose sugars into their ethyl glycosides. It was indicated that the conversion of furfural from pentoses is decreased and ethyl glycosides are more stable than the parent pentoses. Moreover, the anomeric component of the products was rational with a potent transglycosylation reaction mechanism than the hydrolysis followed by glycosylation and ethyl glycosides are potential high value intermediates chemicals. Gandolfi et al.[238] investigated the fractionation of hemp hurds into three main components such as cellulose, hemicellulose, and lignin using organosolv pretreatment. The influence of different process parameters such as temperature, catalyst concentration, reaction duration, and methanol concentration, on the dissolution and recovery of hemicellulose and lignin was studied. Around 75% of total hemicellulose and 75% total lignin were removed using 45% methanol and 3% H_2SO_4 loading at 165°C, 20 min process duration with a little amounts degradation products and 60% of cellulose to glucose yield was achieved through enzymatic hydrolysis of the pretreated biomass. Kabir et al.[239] studied the organoslov pretreatment of forest residues using acetic acid, ethanol, and methanol, and the accumulated methane yields between 0.23 and 0.34 m^3/kg volatile solid were achieved using anaerobic digestion of the forest residues treated at 190°C with 50% (v/v) organic solvent for 60 min pretreatment time whereas 0.053 m^3 CH_4/kg volatile solid was obtained from untreated samples. Chen et al.[240] explored the organosolv pretreatment of wheat straw using formiline, acetoline, sulfuric acid catalyzed ethanol and autocatalyzed ethanol and comparative studies were carried out in terms of the enzymatic digestibility of cellulose, simultaneous saccharification and fermentation (SSF) for ethanol production. It was found that formaline and acetoline favored a higher rate of delignification, lower xylose degradation and solid glucan yields than those obtained from sulfuric acid-catalyzed ethanol and auto catalyzed ethanol pretreated samples. Nitsos et al.[241] investigated the organosolv pretreatment of hardwood and softwood biomass samples using ethanol and sulfuric acid as a catalyst and different process parameters such as ethanol content, catalyst loading, particle size, and pretreatment duration and properties of the isolated lignin were studied. Recovered lignin exhibited high purity above 93% in spite of the more extensive biomass dissolution into process medium and they also showed a lower range of molecular weights. Phenolic hydroxyl component was enhanced due to the lower molecular weight and it was reached to 4 mmol/g with simultaneous decrease in aliphatic hydroxyl component as low as 0.6 mmol/g. Finally, 62% and 69% effective lignin dissolutions were achieved using pretreated spruce and birch respectively along with extensive hemicellulose removal. Recently, Asadi et al.[242] carried out the ethanol organosolv pretreatment of rice straw at various temperature of 120–180°C, ethanol concentration of 45–75% (v/v) and residence time of 30–90 min for synthesis of biohydrogen using Enterobacter aerogenes.

4.6 EXTRACTIVE AMMONIA (EA) PRETREATMENT

The extractive ammonia pretreatment (EA) is the most advanced technique and use of EA pretreatment of lignocellulosic biomass has recently showed much attention in biofuel area. Native cellulose (CI) obtained from lignocellulosic biomass is hardly digestible crystalline nature by fungal cellulases and it was studied that the rate of enzymatic hydrolysis can be enhanced 2 to 5 folds by deconstructing the native cellulose (CI) ultrastructure to another allomorph called cellulose III (CIII).[243,244] Treatment of cellulose with liquid ammonia is used commercially for improving the textile fiber properties at low pressures and sub-zero temperatures.[245,246] In a typical EA pretreatment technique, high ammonia to biomass loading and low moisture content substrate are essentially used to submerge the biomass in liquid ammonia completely for synthesis of an intermediate cellulose ammonia complex and subsequently conversion of complex into CIII followed by ammonia recovery.[247,248] High concentration of water during the process does not favor the synthesis of CIII but it influences the ammonia complex to form native CI.[244,247,249] Synthesis of CIII is very difficult using available ammonia based pretreatment methods such as ammonia fiber expansion (AFEX) and ammonia recycle percolation (ARP) as high moisture content substrate and low ammonia to biomass loading are used in available ammonia pretreatment methods.

Costa Sousa and co-workers[250] investigated an extractive ammonia (EA) pretreatment of corn stover and they reported that lignin extraction from the sample about 45% was achieved with almost preserving the quantitive retention of all polysaccharides and higher fermentable sugar yield was observed using 60% lower enzyme loading than conventional AFEX process. High biofuel yield about 18.2 kg ethanol/100 kg untreated corn stover (dry weight basis) was obtained than that obtained using ionic liquid pretreatments of corn stover, which was very much industrially relevant aspects using low enzyme loading (7.5 mg protein/g glucan) and high solids loading (8% glucan, w/v).[250] It was reported that CIII synthesis is not initiated after ammonia leaves the system at the end of the pretreatment process because the high moisture contents are used near the cellulose fibers during AFEX.[251] On the contrary, for enhancing the reaction temperature heat is supplied to the system from external source and as temperature increases, ammonia pressure builds up until a new vapor-liquid equilibrium is reached. During this reaction course, the cellulose ammonia complex is formed,[252] easter bonds are broken down,[253] and lignin is partially solubilized in the liquid ammonia.[249,253] Likewise AFEX technique, the EA technique favors ammonolysis of cell wall ester cross links that are especially rich in monocots.[253,254] Many researchers reported that EA technique is highly selective towards solubilization of aromatic lignin vs carbohydrate polymers and lignin is the main obstacle to polysaccharide accessibility by biomass degrading enzymes[255] and microorganisms.[256] The development of EA process provides a great significance for future lignocellulose biorefinering processes.

4.7 IONIC LIQUIDS (ILS) PRETREATMENT

The diversity and distinctive properties of ionic liquids have attracted the focus of the scientific interest for several uses such as extraction processes, chemical reactions, electrochemical processes, etc. in recent years. Ionic liquids are considered efficient and "green" novel cellulose solvents since no toxic or explosive gases are formed. Ionic liquids are salts, typically consisted of large organic cations and small inorganic anions, which exists as liquids at normal ambient temperature. Adjustment of the anion and the alkyl constituents of the cation influence the solvent properties such as chemical and thermal stability, non-flammability, low vapor pressures, and a tendency to remain liquid in a wide range of temperatures.[257] Ionic liquids with anion activity favor dissolution of carbohydrates and lignin simultaneously as ionic liquids make hydrogen bonds between the non-hydrated chloride ions of the ionic liquid and the sugar hydroxyl protons in a 1:1 stoichiometry. Therefore, the convoluted chain of non-covalent interactions among biopolymers of lignin, hemicellulose, and cellulose is productively disorganized while minimizing synthesis of degradation products. It was reported that decrystalization of lignocellulosic feed stocks such as straw and wood was investigated efficiently with ionic liquid pretreatment method.[258,259] It was revealed that 54.8% sugar yield upon enzymatic hydrolysis for 12 h is achieved using 1-ethyl-3-methyl imidazolium diethyl phosphate ionic liquid pretreated wheat straw at 130°C for 30 min.[258] It was reported that no negative effect was observed on the growth of *Saccharomycescerevisia* during fermentation of hydrolysates synthesized from enzymatic saccharification of ionic liquid pretreated wheat straw.[258]

The development of cost effective recycling process for ionic liquids, ionic liquid residues and removal of toxicity to enzymes and fermentative microorganisms are essential aspects to be studied extensively before ionic liquids can be considered a real alternative for biomass pretreatment for large scale operation.[260,261] Further, the development of effective recovery techniques for hemicellulose and lignin from solutions after extraction is essential.[257] In spite of these present limitations, advanced study such as promising synthesis of ionic liquids from carbohydrates, may play a significance role in decreasing their cost. Therefore, the application and development of ionic liquids pretreatment offer a great venture for future lignocellulosic biorefinering processes. Shi et al.[262] studied a one-pot, wash-free ionic liquid pretreatment of switchgrass followed by saccharification into a single vessel as conventional ionic liquid process requires excessive amounts of water to remove the ionic liquids from biomass after pretreatment in order to be effective. The pretreated slurry was subsequently hydrolyzed using a thermostable ionic liquid tolerant enzyme cocktile just after treating the switch grass using [C$_2$mim][OAc] ionic liquid and followed by dilution with water to a final ionic liquid concentration of 10–20%. Around 81.2% glucose and 87.4% xylose were achieved using 10%[C$_2$mim][OAc] at 70°C and an enzyme loading of 5.75 mg/ g of biomass and separation of glucose and xylose were carried out using liquid-liquid extraction with above 90% efficiency and thus avoiding extensive water washing. Verdia et al.[263] reported the application of the protic ionic liquid 1-butylimidazolium hydrogen sulfate in the depolymerization and fractionation of lignocellulosic biomass and effects of solution acidity on the depolymerization of Miscanthus giganteus was studied at various 1-butylimidazole to sulfuric acid ratio. Process duration can be reduced using increased acidity and consequently, hemicellulose content can be decreased. Moreover, saccharification yields can be increased by adding more water to the ionic liquid. George et al.[264] synthesized a number of ionic liquids for pretreatment of switch grass with focusing the reduction of solvent cost and stability of the ionic liquid during process. Allison et al.[265] carried out the pretreatment of tomato pomace using the ionic liquid 1-ethyl-3-methylimidazolium acetate ([C$_2$mim][OAc](BASF, Ludwigshafen, Germany)) at different temperatures of 100, 130 and 160°C for 1, 2 and 3 h pretreatment time. The enzymatic digestibility of pretreated sample was carried out using Trichoderma reesei (Sigma Aldrich, St. Louis, MO, USA) at 45°C. Sugar yield was enhanced potentially using certain pretreatment coditions along with significant reduction of pretreatment time. Uju et al.[266] investigated the peracetic acid pretreatment of bagasse and seaweed waste with ionic liquid (IL)-HCl hydrolysis. The cellulose conversion from 20% to 70% was improved using the peracetic acid pretreated bagasse prior to its 1-buthyl-3-methylimidazolium chloride ([Bmim][Cl])-HCl hydrolysis in 1.5 h. Recently, Chang et al.[267] studied an environmentally friendly method for pretreating rice straw with 1-Allyl-3-methylimidazolium chloride ([AMIM]Cl) as an ionic liquid (IL) assisted by surfactants and influence of different surfactant type such as nonionic-, anionic-, cationic- and bio-surfactant were studied. Kassaye et al.[268] investigated the valorization of bamboo biomass regenerated from alkaline solution and ionic liquid pretreatment processes and followed by dilute sulphuric acid hydrolysis.

Parthasarathi et al.[269] reported an effective pretreatment of switch grass using the comparatively low-cost ionic liquid containing tetrabutylammonium [TBA]$^{(+)}$ and hydroxide [OH]$^{(-)}$ ions and it was found that higher glucose yields of about 93% were achieved using ionic liquid pretreated sample at 50°C for 3 h whereas 72.2% glucose yields were found using pretreated sample at 25°C for 3 h. Moreover, glycome profiling experiments and computational results revealed that removal of the noncellulosic polysaccharides occurred due to the ionic mobility of [TBA][OH] and process modeling and energy demand analysis indicated that this [TBA][OH] pretreatment can significantly lower the energy required in the process operation by more than 75%.

4.8 MILD REDUCTIVE CATALYTIC PRETREATMENT (MRCP)

The acceptance of catalytic hydrogenolysis as a valuable approach to the study of lignin structure through the production of significant lignin degradation products efforts have focused on mild reductive catalytic pretreatment.[270] Pepper and Lee[270] reported the detailed comparative study of the effectiveness of various catalysts for the hydrogenolysis of spruce wood lignin. The catalysts studied were Raney nickel, 10% palladium-charcoal, 5% rhodium-charcoal, 5% rhodium-alumina, 5% rutheniumi-harcoal, and 5% ruthenium-alumina. Lignin degradation products were obtained initially as a chloroforn soluble fraction which was then divided and studied as diethyl ether soluble and insoluble fractions. Gas-liquid chromatographic separation of the ether soluble fraction made possible the characterization and quantitative estimation of many of the lower molecular weight lignin degradation products. The data indicated that rhodium, palladium, and a limited amount of raney nickel produce similar results as do ruthenium and an excess of raney nickel; however, with the latter catalysts the degradation is more severe. In particular, rhodium-charcoal and palladium-charcoal appear to offer interesting advantages as catalysts for lignin hydrogenolysis.

4.9 REDUCTIVE CATALYTIC FRACTIONATION (RCF)

In the pioneer second generation ethanol biorefineries being constructed and brought online over the last several years, the residual lignin following saccharification and ethanol fermentation is routed for combustion to produce heat and power or for co-firing with coal, leading to poor valuation of lignin in lignocellulosic biorefining. Indeed, most pretreatment and fractionation techniques being scaled today yield lignin that is even more recalcitrant than native lignin, as pretreatment often relies on the use of mineral acids or hot water, both of which can readily cleave C-O bonds to form reactive intermediates that condense to more refractory C-C bonds.[271–273] Driven by the critical need to improve biorefinery economics and sustainability, conversion strategies aimed at upgrading lignin into biofuels and biochemicals have recently experienced a substantial resurgence. New catalytic and biological approaches offer promise to overcome the primary barriers found in lignin valorization.[274–277] Of particular interest is the work from Pepper and Lee,[270] demonstrating that a mild reductive catalytic treatment, dubbed reductive catalytic fractionation (RCF) by Schutyser et al.,[278] can generate a narrow set of lignin-derived products at high yields (often > 40%) from hardwoods.[270,279–283] Most RCF studies employ late transition metal catalysts with hydrogen gas or a hydrogen donor (e.g., methanol, isopropanol, or formic acid) to cleave aryl-ether bonds and stabilize the resulting reactive species via reductive pathways. Subsequent studies have demonstrated that parameters such as catalyst type, biomass feedstock, solvent, and reaction conditions, profoundly impact product yield, selectivity and distribution.[284] Two common observations across different studies converge on the critical roles of the solvent in rapidly cleaving the abundant β-O-4 bonds in lignin at mild temperatures (~200–250°C) and of the heterogeneous catalyst in stabilizing fragments even without being in direct contact with the solid biomass. RCF is very effective on hardwoods because these feed stocks typically exhibit high syringyl-to-guaiacyl (S/G) monolignol ratios, which translate into high proportions of easily cleavable β-O-4 linkages relative to the more refractory C-C linkages. In contrast, feed stocks with lower S/G ratios, such as softwoods or herbaceous biomass, feature a higher proportion of C-C linkages, making RCF more challenging. In particular, herbaceous feed stocks exhibit low S/G ratios (0.62) and higher content of the hydroxycinnamic acids (HCAs), pcoumaric and ferulic acids.[285–289] The p-coumaric acid moieties are typically pendant to the syringol moieties in lignin, whereas ferulic acid polymerizes through a broad suite of linkages both internal and pendant to the lignin polymer, as well as via covalent attachment to hemicellulose. Schutyser et al.[278] reported lignin monomer yields of ca. 50%, 21%, and 27% after the RCF of birch (a hardwood), pine/spruce (a softwood mixture), and miscanthus (a grass), respectively.[279,290] Indeed, herbaceous feed stocks such as corn stover, switch grass, wheat straw, sugarcane bagasse, and miscanthus represent a large fraction of the biomass available in the global bioeconomy.[291] As such, new RCF approaches for herbaceous feed stocks need to be developed to selectively activate a more diverse suite of linkages for effectivedelignification and subsequent upgrading of the resulting stream of lignin fragments. Eric Michael Anderson et al.[292] investigated recently the RCF of corn stover an abundant feedstock in lignocellulose conversion, especially for North America.[293] Corn stover lignin is composed of up to ~20% pcoumaric acid and ~10% ferulic acid and a low S/G ratio of 0.62.[288,289] They presented a comprehensive study on the role of reaction conditions, catalyst type, mineral acid co-catalysts, and the acidity of the catalyst support to gain insight into the underpinnings of solvolysis and fragment stabilization. The use of mineral acid co-catalysts has been demonstrated for hardwood lignin, but has not been explored under reductive conditions for the extraction of herbaceous biomass that contains a considerable amount of ester linkages.[294] Using carbon-supported Ru and Ni catalysts at 200 and 250°C in methanol and, in the presence or absence of an acid co-catalyst (H_3PO_4 or an acidified carbon support), three key performance variables were studied:1) the effectiveness of lignin extraction as measured by the yield of lignin oil, 2) the yield of monomers in the lignin oil, and 3) the carbohydrate retention in the residual solids after RCF. The monomers included methyl coumarate/ferulate, propyl guaiacol/syringol, and ethyl guaiacol/syringol. The Ru and Ni catalysts performed similarly in terms of product distribution and monomer yields. The monomer yields increased monotonically as a function of time for both temperatures. At 6 h, monomer yields of 27.2 and 28.3% were obtained at 250°C and 200°C, respectively, with Ni/C. The addition of an acid co-catalysts to the Ni/C system increased monomer yields to 32% for acidified carbon and 38% for phosphoric acid at 200°C. The monomer product distribution was dominated by methyl coumarate regardless of the use of the acid co-catalysts. The use of phosphoric acid at 200°C or the high temperature condition without acid resulted in complete lignin extraction and partial sugar solubilization (up to 50%) thereby generating lignin oil yields that exceeded the theoretical limit. In contrast, using either Ni/C or Ni on acidified carbon at 200°C resulted in moderate lignin oil yields of ca. 55%, with sugar retention values greater than 90%.These sugars were amenable to enzymatic digestion, reaching conversions greater than 90% at 96 h. Characterization studies on the lignin oils using 2D HSQC NMR and GPC revealed that soluble oligomers are formed via solvolysis, followed by further fragmentation on the catalyst surface via hydrogenolysis. Overall, the results show that clear tradeoffs exist between the levels of lignin extraction, monomer yields, and carbohydrate retention in the residual solids for different RCF conditions of corn stover.[292]

4.10 DEEP EUTECTIC SOLVENTS METHODS

Deep eutectic solvents (DESs) are a green and efficient alternative to ILs for biomass pretreatment and conversion. DESs are a special mixture of two components: hydrogen-bonding donor (HBD) and hydrogen-boning acceptor (HBA).[295] The components of DESs could be extended into three or more components. The melting point of DESs is much lower than its components (HBD and HBA) due to the strong hydrogen-bonding interaction between HBD and HBA. The DESs are easy to prepare, carry a low cost, have low toxicity and high biodegradability. They have attracted much attention in many fields.[296–299] Particularly, these intramolecular hydrogen bonds in DESs render the high possibility of breaking strong hydrogen bonds among biomass, hence the high biomass solubility and the favorable conversion rate.[300–305]

Guo et al. measured the solubility of cellulose in several DESs and obtained the order: ChCl:imidazolium (ChCl:IM, 2.48 wt%) >ChCl:urea (1.45 wt%)>ChCl: ammonium thio-cyanate (ChCl:AT, 0.83 wt%).[306] Both ChCl:IM and ChCl:urea owned a higher solubility than benchmark ILs 1-allyl 3-methylimidazolium chloride ([AMIM][Cl], 1.89 wt%). The cellulose solubility correlated well with the hydrogen-bonding basicity (b) of DESs and ILs: ChCl:IM (0.864)>[AMIM][Cl] (0.830)>ChCl:urea (0.821)>ChCl:AT(0.810). However, the correlation between cellulose solubility and dipolarity/polarizability was not favorable. It means that the hydrogen-bonding basicity of DESs determined the solubility of cellulose in DESs. A higher value of hydrogen-bonding basicity favored an easier breakdown of hydrogen bonds in the supermolecular structure, hence a higher solubility of cellulose (Table 4.1).[307]

Besides being used for the dissolution of cellulose, DESs have been applied for assistance transformation of cellulose into cellulose nanocrystals, and the obtained cellulose nanocrystals can be used as environmentally safe dispersing agents to stabilize oil in water emulsions.[308] The yield (74.2%), the time (3 min via ultrasonication),

TABLE 4.1
Solubility of the Isolated Biomass Components in Various DESs at 60°C

Hydrogen bond donor	Hydrogen bond acceptor	Mole ratio	Lignin mass solubility (%)	Cellulose mass solubility (%)	Xylin mass solubility (%)
Formic acid	ChCl	2:1	14	<1	<1
Lactic acid	ChCl	10:1	13	<3	<5
Acetic acid	ChCl	2:1	12	<1	<1
Lactic acid	Betaine	2:1	9	<1	<1
Lactic acid	Proline	3.3:1	9	<1	<1

the size (3–25 nm), and the crystallinity (82%) of cellulose nano-crystals could be improved by DESs from cotton after breaking the strong hydrogen-bonding of cotton.[309] The efficiency via ultra-sonication was higher than that via a strong acid or base.

5

BIOLOGICAL PRETREATMENT

In biological pretreatment, wood degrading microorganisms such as white-, brown-, soft-rot fungi, and bacteria are utilized to change the chemical composition and/or structure of the lignocellulosic biomass so that the reconstructed biomass is more susceptible to enzyme digestion. The frequently used microorganisms for biological pretreatment are white-rot fungi that belong to class of Basidiomycetes.[310] Biological pretreatment provides advantages such as low chemical and energy requirements, safe and environment friendly though adequate rapid and controllable system is under study yet.[311]

5.1 DEGRADATION OF CELLULOSES

The synthesis of celluloses was increased by the bio agents through the degradation of sugarcane trash and around 12, 10, 9, and 8 folds higher degradation were observed using *Aspergillus Terreus, Terreus, Cellulomonas Cellulomonas Uda, Cellulomonas Cartae* and *Bacillus Macerans,* respectively. It was reported that bio-organosolvation pretreatments of beech wood chips were studied using ethanolysis and four white-rot fungi such as *Ceriporiopsis sub Vermispora, Dichomitus Squalens, Pleurotus Ostreatus,* and *Coriolus Versicolor* without addition of any nutrients for 2 to 8 weeks, and pulp and soluble fractions were separated by ethanolysis method followed by simultaneous saccharification and fermentation (SSF) of pulp portion.[312] The highest yields of ethanol 0.294 g/g of ethanolysis pulp (74% of theoretical) and 0.176 g/g of beech wood chips (62% of theoretical) were achieved using solid-state fermentation (SSF) of *C. sub Vermispora* pretreated samples for 8 weeks, and the yield was 1.6 fold higher than that synthesized without the fungal treatments, and biological pretreatments saved 15% of the energy needed for ethanolysis. Balan and his co-workers[313] carried out the biological pretreatment of rice straw with white-rot fungi *Pleurotus Ostreatus,* followed by AFEX process, and they reported

DOI: 10.1201/9781003203414-5

that higher conversion of glucan and xylan was observed using enzymatic hydrolysis of biologically pretreated samples followed by AFEX process than that obtained from rice straw with AFEX process solely. Low energy requirements and moderate process conditions are the advantages of the biological pretreatment method, although the low hydrolysis rate of most biological pretreatment processes is the major limitation along with solubilization and consumption of lignin, hemicellulose, and cellulose by most lignolyte microorganisms, and due to above limitations the biological pretreatment faces technoeconomic challenges and is less attractive commercially. Zhao et al.[314] studied the fungal pretreatment of cornstalk using *Phanerochaete Chrysosporium* for enzymatic saccharification and H_2 synthesis, and about 34% delignification and 10% hemicellulose were achieved at 29°C for 15 days under static condition operation. The higher sugar yield of 47.3% was found from enzymatic hydrolysis of the fungal pretreated cornstalk with crude cellulase from Trichoderma viride than that of 20.3% obtained from untreated sample, and H_2 yield of 80.3 ml/g-pretreated cornstalk was achieved from fermentation of the sugars with *Thermoanaerobacterium Thermosaccharolyticum* W16. It was reported that their results revealed the potential of using hydrogen-producing bacteria for high-yield conversion of cornstalk into bio-H_2 integrate with biological pretreatment technique and enzymatic saccharification. Liu et al.[315] investigated the fungal pretreatment of switch grass involving SSF to increase the saccharification and simultaneously produce enzymes as co-products, and it was observed that 30% delignification was achieved with a considerable amount of *laccase*, as high as 6.3 U/g using the fungus *Pycnoporus sp.* SYBC-L3 for 36 days cultivation period without potential loss of cellulose and hemicellulose. Finally, about 50% higher hydrolysis yield was observed with 36 days pretreatment period than that obtained from untreated switch grass. Cianchetta et al.[316] carried out the biological pretreatment of wheat straw using fungi for a low cost and eco-friendly technique to physicochemical methods to facilitate enzymatic hydrolysis. They conducted the study with five various fungi on enzymatic hydrolysis of wheat straw, and the best sugar yield of around 44% was synthesized using a *Ceriporiopsis Subvermispora* strain for 10-week pretreatment period, which was more than double of the yields obtained with other isolates, and the hemicelluloses in the pretreated biomass exhibited an inverse relation with digestibility. Wen et al.[317] studied the biological pretreatment of Napier grass using three microbial consortia such as MC1 (*Clostridium straminisolvens*), WSD-5(*Coprinus cinereus* and *Ochrobactrum sp.*), and XDC-2 (mesophilic bacteria in the genera *Clostridium*, *Bacteroides*, *Alcaligenes*, and *Pseudomonas*) followed by saccharification and anaerobic digestion. Around 66.2% and 43.4% total sugar yields were achieved with moderate (Cellulase:b-glucosidase: xylanase::40:4:4(IU/g of substrate dry basis)) and low (Cellulase:b-glucosidase: xylanase::15:1.5:1.5 (IU/g of substrate dry basis)), respectively, using consortium WSD-5 pretreated samples, whereas the highest sugar yield of 83.2% was found with high enzyme loading (Cellulase:b-glucosidase: xylanase::70:7:7 (IU/g of substrate dry basis)) using consortium MC1 pretreated samples at 50°C for 48 h hydrolysis duration. The maximum methane yield from pretreated samples using the consortia MC1, WSD-5, and XDC-2 were 259, 279, and 247 ml/g volatile solid, respectively, which were 1.39, 1.49, and 1.32 folds higher than the values obtained from the untreated Napier grass. Kumari and Das[318] reported an eco-friendly biological pretreatment of sugarcane top for efficient delignification, and it was found that 60.4% (w/w) lignin removal was achieved at 28°C after the incubation of 21 days in static condition. The highest hydrogen production of 16.76 mL/g-volatile solid/h was observed using pretreated samples and highest methane synthesis of 180.86 Ml/g volatile solid was found from spent medium of dark fermentation. Kavitha et al.[319] investigated the effect of bacterial-based biological pretreatment on liquefaction of the microalga Chlorella vulgaris prior to anaerobic biodegradation to gain insights into energy-efficient biomethanation. Liquefaction of microalgae resulted in a higher biomass stress index of around 18% in the experimental samples (pretreated with cellulose-secreting bacteria) versus 11.8% in the untreated samples. Mathematical modeling of the biomethanation study revealed that bacterial pretreatment facilitated sustainable methane recovery, with a methane yield of around 0.08 (g chemical oxygen demand/g chemical oxygen demand), compared to the yield of 0.04 (g chemical oxygen demand/g chemical oxygen demand) obtained from control pretreatment.

5.2 DEGRADATION OF LIGNIN AND HEMICELLULOSE

Degradation of lignin and hemicelluloses in waste materials is accomplished using biological pretreatment with microorganisms such as brown, white, and soft rot fungi.[13] It was studied that brown rots mostly attack cellulose crystal, while white and soft rots attack both cellulose and lignin, and lignin degradation using white-rot fungi takes place under the action of lignin degrading enzymes such as peroxidases and lactase, which are governed by carbon and nitrogen nutrients sources.[320] It was also reported that lignocellulosic biomass compounds are pretreated efficiently using white-rot fungi.[30] It was revealed that 35% sugar yield is achieved using a biologically pretreated straw with *Pleurotus Ostreatus,* and exactly the same performance is observed in the pretreatment of straw by *Phanerochaete Sordida*[105] and *Pycnoporus Cinnabarinus*[311] in 4 weeks.[321] It was published that delignification was developed using biologically pretreated wood chips with a cellulose-less mutant of *Sporotrichum pul Verulentum.*[322] Akin and co-workers[323] reported that 32% and 77% biodegradation of Bermuda grass stems was noticed using white-rot fungi such as *Ceriporiopsis subVermispora* and *Cyathus Stercoreus,* respectively, in 6 weeks. Lee and co-workers[320] carried out the biological pretreatment on the Japanese red pine *Pinus Densiflora* using three white-rot fungi such as *Ceriporia Lacerata, Stereum Hirsutum,* and *Polyporus Brumalis,* and it was found that total weight loss was 10.7% using *S. Hirsutum* fungus in 8 weeks and lignin loss was 14.5% which was the highest among the tested samples, but the hemicelluloses loss was 17.8% which was lesser than the losses observed using *C. Lacerata* and *P. Brumalis* fungi. On the contrary, extracellular enzymes from *S. Hirsutum* exhibited best activity of ligninase and lowest activity of cellulose than those from white-rot fungi, and finally, *S. Hirsutum* was selected as an effective promising fungus for biological pretreatment. It was also published that a 21% higher sugar yield was achieved using enzymatic saccharification (commercial enzymes, Celluclast 1.5 L and Novozyme 188) of *S. Hirsutum*-treated Japanese red pine than untreated samples. It was reported that lignin degrading enzymes such as lignin peroxidases and manganese dependent peroxidases were produced from white-rot fungus *P. Chrysosporium* during secondary metabolism, in response to carbon or nitrogen limitation.[324] Both the enzymes were obtained in the extracellular filtrates of some white-rot fungi for rupturing the wood cell walls.[325,326] Bazhal et al.[327] studied the biological pretreatment of sugarcane trash using some fungi and bacteria at different carbon to nitrogen (C/N) ratio of trash from 108:1 to a varying range from 42.1 to 60:1, and it was reported that 61%, 52%, and 49% decrease in C/N ratios were noticed using

Aspergillus Terreus,Cellulomonas uda and *Trichoderma reesei,* and *Zymomonas mobiliz,* respectively. They inferred that C/N ratio is an important parameter for lignocellulosic biomass pretreatment as dissociation of lignocellulosic biomass depends on the C/N ratio of the biomass and indicated that a definite fraction of nitrogen is essential for microorganisms to degrade each molecule of carbon, and this fraction changes with various kinds of microflora. It was reported that fungi are more effective to degrade any lignocellulosic biomass material than bacteria as fungi have a higher C/N ratio of 30:1 as compared to 10:1 for the bacteria, as their dependency on nitrogen is comparatively lower.[320]

6

COMBINED PRETREATMENT

Pretreatment of lignocellulosic biomass with more than one pretreatment technique subsequently favors the sugar yields significantly. It was reported that delignification and rate of the enzymatic hydrolysis were enhanced using photocatalyst-assisted alkali pretreated rice straw. Zhu and his co-workers[328] studied the microwave-assisted acid and alkali pretreatments of rice straw and they reported that higher hemicellulose and lignin removal was achieved using microwave-assisted acid and alkali pretreated samples than that of acid and alkali pretreated rice straw. It was revealed that higher microwave power requires short pretreatment time and lower microwave power requires longer pretreatment time. Lu and Minoru[329] studied the electron beam irradiation of rice straw in NaOH solutions and it was observed that the delignification was enhanced using the electron beam irradiation as it modified the lignocellulosic structure which facilitated higher accessibility of NaOH into the lignocellulosic biomass and finally higher rate of enzymatic hydrolysis was noticed. It was reported that combined steam explosion and super fine grinding pretreatment of rice straw was investigated at low sever conditions and higher enzymatic hydrolysis rate was reported along with short grinding time, low energy cost, and absence of inhibitors.[330] To avoid excessive decomposition of hemicellulose and side products synthesis from sugars and lignin super fine grind was carried out just after the rice straw was pretreated with steam explosion. Higher sugar yield was observed using the superfine ground samples than that of untreated rice straw and interestingly, the reducing sugar yield obtained from the superfine ground residue was even lower than that obtained from untreated rice straw. Steam explosion followed by superfine grinding favor to decrease in particle size and increases active surface area for higher accessibility of enzymes than that of the conventional mechanical grinding. Zhu et al.[331] studied the microwave-assisted NaOH and H_2SO_4 pretreatment of miscanthus at different temperatures (130 to 200 °C) for 20 min. Significance sugar yield ca. 73% was achieved at

DOI: 10.1201/9781003203414-6

temperature 180°C with 0.2 M H_2SO_4 and sugar yield was increasing with increasing temperature using pretreated miscanthus, which was 17 times higher within half the time than that obtained from conventional heating pretreated miscanthus. Verma et al.[332] carried out the combined pretreatment of beech wood with microwave-assisted irradiation and nine different ammonium salts in the presence of H_2O_2. The highest sugar yield of ca. 60% was reported from microwave irradiated at 140°C for 30 min using ammonium molybdate salt while ca. 42% sugar yield was obtained from the external heating sample in an autoclave. They also indicated that an ammonium ion is the key counterion accelerating the pretreatment with molybdate and their study highlights the potential selective delignifying capability of the H_2O_2-activated ammonium molybdate system assisted by microwave radiation. Ethaib et al.[333] studied the microwave-assisted diluted 0.1 N H_2SO_4, 0.1 N NaOH, and 0.01 N $NaHCO_3$ solvents pretreatment of sago palm bark, and highest degradation of hemicellulose was achieved from 0.1 N H_2SO_4 treated sample and maximum glucose (9 mg/g) and xylose (4 mg/g) yields were found using 0.1 N H_2SO_4 pretreatment and acetic acid was indicated as an inhibitors in 0.1 N H_2SO_4 pretreatment. Bala et al.[334] studied the combined ultrasound-assisted alkaline pretreatment of cellulose Iβ and it was observed that combined pretreatment favors a great rupture of cellulose particles along with the formation of large pores and partial fibrillation. The influences of ultrasound irradiation time (2, 4 h), NaOH concentration (1–10 wt%), initial particle size (20–180 μm), and initial degree of polymerization (DP) of cellulose on structural modification and glucose yields were studied. The alkaline ultrasonic pretreatment favored a significant reduction of the particle size of cellulose and a decrease in the treatment time and concentration of NaOH. The glucose yields showed an increasing propensity with an increase in cellulose II fraction as an outcome of combined pretreatment and above 2.5 fold enhancement in glucose yield was achieved using 10 wt%NaOH for 4 h of sonification, compared to untreated samples and no influence DP on glucose yield was observed.

7

PROBLEMS OF THE INDUSTRIAL ADAPTATION

7.1 ENERGY

The biofuel sector is both a consumer and supplier of energy. As a supplier of energy, biofuels replace oil or electricity while they consume coal, natural gas, and electricity during production. The impact of these commodities on price at least in the near term should be minimal if biofuel production remains at low levels compared to the world or regional demand for those commodities. (Of course, in the electricity sector cost of marginal supply determines the market-clearing price; but the same is not true of oil, coal, and gas markets since these goods are storable.) However, a scenario in which a major oil-consuming region such as the United States or the European Union (EU) meets a significant portion of its demand from biofuel can cause a reduction in world oil prices. That said, environmental regulations like carbon taxes may dwarf the effect of biofuels on the price of fossil fuels. However, this effect has not been investigated by past studies.

7.2 FOOD

Biofuels will increase the price of food either because food crops are converted to fuel or because energy crops replace food crops on agricultural lands. The ultimate impact on a region will depend on several factors including the intensity of cultivation of biofuel crops and the extent of trade in food-related commodities. One can envision several scenarios. Developed regions such as the EU and the United States will experience price increase but may be able to absorb the price rise more easily than developed countries. One reason for this could be that since food-processing costs comprise a large share of the total cost, there will be a lesser impact on the final consumer price. The food processing industry will, however, be

   DOI: 10.1201/9781003203414-7

negatively affected due to higher input costs and lower demand for food. Developing countries that are net importers of food will be negatively affected due to higher food prices irrespective of whether they adopt biofuels or not.

7.3 LIVESTOCK

Biofuels will have a mixed effect on the livestock sector. When crops used for feed are diverted to ethanol, higher feed prices and higher livestock prices will result. The US Department of Agriculture also predicts higher price of crops such as soy, sorghum, alfalfa, and hay, which are displaced due to greater planting of corn to meet biofuel demand. However, higher feed prices will be partially offset by the increased supply of dried distillers grains with solubles, which is a coproduct of ethanol production and is a substitute for corn. The response by the livestock sector to changes in feed prices depends on the relative importance of protein (primarily soybean meal) versus energy (primarily corn) and the size of the price changes associated with these feed components.

7.4 LAND

Allocating land to biofuels means taking land away from other uses such as food or environmental preservation. The demand for agricultural land will benefit landowners. In fact, it has been hypothesized that tenant farmers may end up losing much of the benefits to landowners in the form of increased rent. Increased demand for land for farming could lead to expansion of the agricultural land base. This might result in marginal and environmentally sensitive lands being brought under production such as CRP lands in the United States, set-aside lands in Europe, and tropical rainforests in Indonesia.[335,336] In other cases biofuel plantations may come up at the expense of pastureland rather than cropland or forestland. It is estimated that in Brazil about 30 million hectares out of the 220 million hectares could migrate to crops with little impact on meat production due to technological advance.[337] The conversion of such lands to crop production will release carbon sequestered in the soil into the atmosphere, which will offset some of the carbon benefits. One concern that has been raised by critics of biofuels in this regard is that governments tend to reclassify forestlands as pasturelands in order to facilitate their conversion to biofuel plantations. In India, biofuel plantations are planned for wastelands, which are considered not suited for cultivation of food crops. Here again, the categorization of such lands as wastelands has been disputed given the dependence of the rural poor on those lands for grazing and fuelwood collection.[338,339]

7.5 WATER

Agricultural water demand will also increase either due to the expansion of agriculture or if biofuel crops are more water intensive than traditional crops. The increased demand for water will lead to higher optimal price. This might lead to reduced availability of water for food crops lowering yield and affecting food supply. Because the demand for food is inelastic, the demand for water for food production is inelastic.

7.6 LABOR

Biofuels are more labor intensive than other energy technologies per unit of energy-delivered basis. Therefore, biofuels should result in a net creation of new jobs related to energy production with the bulk of the increase occurring in the agricultural and processing phase. Of course, there will be a reduction in the rate of increase that would have otherwise occurred in the oil processing industry to meet future demand. There will also be movement of labor within the agricultural and the food processing sector.[340]

7.7 FARM INCOME

The impact on farm income will depend on several factors. In general, net food producers will benefit from an increase in food prices. However, energy is also an input to agriculture and, to the extent that energy prices are a significant part of the costs of production, it will dampen the net increase in profits. Similarly, an increase in water prices can also affect the productivity and income of farmers. In places like the EU, where farmers are already granted higher prices than market prices, it seems unlikely that the farmers will be granted an even higher price as a result of the competition from biofuels. Such farmers will not experience an additional

increase in rent. In the absence of such supports, however, the vo-
latility of the energy situation will mean a risk of major losses from
shocks that may lower energy prices.

7.8 AGRIBUSINESS

Modern biofuels have major implications for agribusiness. The
conversion of energy crops into modern biofuels requires sophisti-
cated processing technology. If energy prices are expected to remain
high or if subsidies are likely to remain in place, there will be major
investments made in crop production and processing. The result is
likely to be strategic alliances between farmer cooperative agribusiness and energy sector merger and acquisition of agribusiness by
energy firms.

7.9 INTERNATIONAL TRADE

Biofuels are expected to reduce a nation's dependence on imports for
oil. While this is likely to be true, it, however, may not reduce the
dependence on imports for energy unless the biofuel feedstock is
produced locally. This has led some to question the effectiveness of
biofuels in improving energy security. Simulations for the EU and
the United States suggest that these regions are likely to become net
importers for agricultural commodities especially in scenarios at
levels of substitution.[341,342] Thus, they could have a negative impact
on the balance of trade, and at the same time, developing countries
will experience a reduction in imports and an increase in exports.

8

ECONOMICS OF DIFFERENT PRETREATMENT TECHNOLOGIES

The economic advantage of biofuels is that they are a convenient, low-cost, domestically producible substitute for oil, a fuel that is getting costlier by the day and is also imported from politically volatile regions. The increased demand for agriculture from biofuels can also address the worldwide problem of declining farm income. But negative effects on food and the environment are threatening to offset the positive effects on welfare as an energy source. This, however, should not be surprising. As the previous chapter explained, biofuels are intensive in the use of inputs, which include land, water, crops, and fossil energy, all of which have an opportunity cost. Understanding how biofuels will affect resource allocation, energy and food prices, technology adoption, income distribution, and so forth is essential at this very early stage of development. A variety of economic modeling techniques are being used to model the impacts from different angles.

8.1 ECONOMIC STUDIES OF BIOFUELS

The biofuel production chain can be divided into the following four stages:

1. **Production of biomass feedstock through cultivation:** This is mainly an agricultural activity in which a biofuel crop is grown, harvested, and transported to a conversion facility. The biofuel crop can be a food crop like corn or a dedicated energy crop like switchgrass.
2. **Conversion of the feedstock to fuel (or electricity):** This is an industrial activity in which the raw biomass is converted into biofuel along with one or more co-products.

   DOI: 10.1201/9781003203414-8

3. **Distribution and retailing of finished fuels:** This involves the distribution of finished fuel for blending with fossil fuels. In the case of electricity, this involves the transmission and distribution of electricity to demand centers.

4. **Consumption of bioenergy:** This refers to the ultimate end use in which the biofuel enters the fuel tank of a vehicle or provides electricity.

From a private standpoint, a variety of economic questions arise at each stage, which relates to if and when a producer or consumer will adopt biofuel. From a societal standpoint, questions like what incentives are needed and what are the aggregate impacts of biofuels on welfare are the most important. We discuss the existing economic literature under four broad categories. These are as follows:

1. Cost accounting models
2. Micromodels of technology adoption and resource allocation
3. Sector models
4. General equilibrium and international trade models.

8.2 ESTIMATES OF FUTURE POTENTIALS FOR BIOENERGY

There are several studies that estimate the global potential of biofuels in absolute units of energy and as percentages of global energy that they can supply. Estimates of such potential can be classified into three categories: namely, biophysical, technical, and economic. Each category in the list comprises the ones following it so that the three categories are of decreasing magnitude. Biofuels can in principle supply a large fraction of global energy needs, and this is called the theoretical potential. The biophysical potential is determined primarily by natural conditions and describes the amount of biomatter that could be harvested at a given time. However, oil prices are uncertain with respect to time, while policies vary both with time and also from region to region.[343] As a result, economic potential is hard to predict. For example, Brazilian ethanol is economically viable when oil sells at $35 per barrel, whereas US ethanol is viable only at around $50 per barrel.[344,345]

Most studies report an increase in the supply of bioenergy over time. A review of earlier studies on this subject by Berndes, Hoogwijk, and van den Broekreveals that estimates for potential contribution of biomass in the year 2050 range from below 100 EJ/yr to over 400 EJ/yr.[346] In comparison to the current level of bioenergy of 45 EJ/yr, this represents a doubling to a tenfold increase. A study by the International Institute of Applied Systems Analysis and the World Energy Council predicts that bioenergy would supply 15% of global primary energy by 2050.[343] A study by the Natural Resources Defense Council predicts that an aggressive plan to develop cellulosic biofuels between now and 2015 could help the United States produce the equivalent of nearly 7.9 million barrels of oil per day by 2050. This is equal to more than 50% of the current total oil use in the transportation sector.[347] A majority of the increase is accounted by cellulosic biomass like switchgrass.

An analysis of the demand for cropland based on fundamental forces responsible for expansion of cropland by Waggoner and Ausubel[348] suggests that sustained technological progress in crop production could meet the recommended nutritional requirements for a population of 9 billion and simultaneously reduce cropland by 200 million hectares by the year 2050. It is even claimed that under the best-case scenario the land withdrawn from agriculture could be as high as 400 million hectares. At the same time, they warn that such improvements would come about only through sustained investments in productivity, experimentation, and deployment of better technologies.[348,349]

8.3 NEW JOBS

Biofuels are more labor intensive than other energy technologies on per unit of energy delivered basis.[350] The production of the feedstock and the conversion requires greater quantities of labor compared to that required for extraction and processing of fossil fuels or other industrially based technologies like hydrogen and electric vehicles. A majority of these job additions are expected to take place in the rural sector which can also spur rural development.[351]

9

ENVIRONMENTAL CONCERNS

9.1 REDUCTION OF CARBON EMISSIONS

Biofuels are sometimes considered as a solution to climate change. While this may be too optimistic, it is true that direct carbon emissions from combustion of biofuels are insignificant compared to fossil fuels. That said, it is hard to generalize about indirect carbon emissions (from agriculture and processing) and emissions of other harmful pollutants, which can be significant.

9.2 IMPROVE ENERGY SECURITY

The aforementioned fact also means that countries can produce their own fuel and reduce their dependence on foreign sources for energy.[352]

9.3 SIMPLE AND FAMILIAR

Finally, biofuels have an aura of being simple and familiar to consumers, producers, and policymakers alike. Ethanol has been used as an additive or as a blend with gasoline in several countries for over two decades. In fact, Henry Ford and Rudolph Diesel who are considered the grandfathers of the automobile assembly line and the diesel engine, respectively, are said to have prophesized a future for transportation based on fuels derived from plant-based sources.[353]

However, if agriculture is to be relied on to fuel a growing population, one that is richer and drives more, then a serious consideration of the consequences of widespread biofuel adoption is warranted; the technology is not without costs. Biofuels may mean filling the fuel tank at the cost of emptying the stomach of the poor.[354,355] Biofuels are also feared for the impact they will have on the natural environment.[354,356–358] Basically, biofuel technology is land intensive. Biofuel demand will put pressure on existing use of land including food production and natural habitats. It will also increase the demand for agricultural inputs like fertilizers, pesticides, and so forth, which have negative environmental externalities. By increasing energy supply, biofuels can also undermine efforts at improving energy efficiency and energy conservation. We defer a more detailed discussion on the environmental and economic implications on biofuels to previous chapters. The emphasis of this chapter is on the sources, technologies, and uses of bioenergy systems.

Traditional biomass accounts for 80% of the global renewable energy use, while ethanol and biodiesel comprise less than 1% of the global renewable energy use (the remaining is accounted for by wind, solar, hydro, geothermal, and tidal energy). In any case, the focus of this survey is largely on liquid biofuels, the reason being that it is one of the fastest-growing sources of alternative energy today. The impacts of the huge investments taking place in developing modern biofuels are not well understood, and hence more controversial, whereas several prominent works on traditional biomass already exists.[359–363]

10

CONCLUSIONS AND
RECOMMENDATIONS

10.1 CONCLUSIONS

The pretreatment methods for lignocellulosic materials have been explained and extensively studied to improve biofuel production processes. All these pretreatment techniques describe the lignocellulosic biomass accessible to enzymatic hydrolysis, where crystalline nature of cellulose, its accessible active surface area, and delignification and hemicelluloses disposal are mostly substrate-associated parameters, which influence the enzymatic hydrolysis. The chemical and thermochemical pretreatment techniques are recently the most efficient techniques among the various methods and include the most potential methods for commercial applications. The combination of various pretreatment methods has been regarded and may be considered to attain optimum fractionation of the various components and achieve very high yields. From primordial research point of view, one approach which is drawing more focus is the study of the influence of pretreatments at a more basic level. Plant cell wall is very complicated and its study at cellular, ultrastructural, and even molecular level can help understand various catalytic reactions acting on biomass as well as the outcomes of the thermal or chemical pretreatments. Finally, it may be inferred that to resolve the technical obstacles of the biomass conversion process, inventive technology and innovative science are to be used, so that biofuel synthesis from lignocellulosic biomass can be developed efficiently.

10.2 CHALLENGES AND FUTURE PERSPECTIVE

In last several decades, biomass pretreatment is used to improve the biodegradability of biomass. Only a few pretreatment methods such as dilute acid and steam explosion have been commercialized for cellulosic ethanol production in spite of carrying out extensive

DOI: 10.1201/9781003203414-10

research to develop numerous effective pretreatment techniques. One of the most famous demonstration scale cellulose ethanol facilities was hosted by Iogen Corporation, Canada. In Iogen Corporation, modified steam explosion pretreatment is used to increase the enzymatic digestibility of straw for production of cellulosic ethanol with the yield of 340 L/ton of fiber. The U.S. Department of Energy sponsors several biofuel programs through bioenergy research institutes such as Great Lakes Bioenergy Research Center, National Renewable Energy Laboratory, and Oak Ridge National Laboratory to improve various pretreatment techniques. The research on lignocellulosic biomass pretreatment presents many technoeconomic challenges such as cellulose depolymerization, delignification, cofermentation of hexose and pentose into ethanol, hydrolyzate detoxification, hemicellulose and cellulose separation, pretreatment product digestibility, inhibitors protection, environmental impact, energy demand, and processing expenditure.

REFERENCES

(1) Yat, S.C.; Berger, A. and Shonnard, D.R. Kinetic Characterization of Dilute Surface Acid Hydrolysis of Timber Varieties and Switchgrass. *Bioresource Technol.* 2008, *99*, 3855–3863.

(2) Weekly U.S. Retail Gasoline Prices, Regular Grade. Energy Information Administration, U.S. Department of Energy: Washington, DC, June, 2008; see http://www.eia.doe.gov/oil_gas/petroleum/data_publications/wrgp/mogas_home_page.html.

(3) Wyman, C.E. Biomass Ethanol: Technical Progress, Opportunities, and Commercial Challenges. *Annu. Rev. Energy Environ.* 1999, *24*, 189–226.

(4) Wang, M.; Wu, M. and Huo, H. Life Cycle Energy and Greenhouse Gas Emission Impacts of Different Corn Ethanol Plant types. *Environ. Res. Lett.* 2007, *2*, 1–13.

(5) Broder, J.D.; Barrier, J.W.; Lee, K.P. and Bulls, M.M. Biofuels System Economics. *World Resour. Rev.* 1995, *7*, 560–569.

(6) Iranmahboob, J.; Nadim, F. and Monemi, S. Optimizing Acid Hydrolysis: A Critical Step for Production of Ethanol from Mixed Wood Chips. *Biomass Bioenerg.* 2002, *22*, 401–404.

(7) Patrick Lee, K.C.; Bulls, M.; Holmes, J. and Barrier, J.W. Hybrid Process for the Conversion of Lignocellulosic Materials. *Appl. Biochem. Biotechnol.* 1997, *66*, 1–23.

(8) Hsu, T.A.; Ladisch, M.R. and Tsao, G.T. Alcohol from Cellulose. *Chem. Technol.* 1980, *10*, 315–319.

(9) Mosier, N.S.; Wyman, C.; Dale, B.; Elander, R.; Lee, Y.Y.; Holtzapple, M. and Ladisch, M.R. Features of Promising Technologies for Pretreatment of Lignocellulosic Biomass. *Bioresour. Technol.* 2005, *96*, 673–686.

(10) Lynd, L.R.; Elander, R.T. and Wyman, C.E. Likely Features and Costs of Mature Biomass Ethanol Technology. *Appl. Biochem. Biotechnol.* 1996, *57*, 741–761.

(11) Lee, J. Biological Conversion of Lignocellulosic Biomass to Ethanol. *J. Biotechnol.* 1997, *56*, 1–24.

(12) Lee, D.; Yu, A.H.C.; Wong, K.K.Y. and Saddler, J.R. Evaluation of the Enzymatic Susceptibility of Cellulosic Substrates Using Specific Hydrolysis Rates and Enzyme Adsorption. *Appl. Biochem. Biotechnol.* 1994, *45*, 407–415.

(13) Galbe, M. and Zacchi, G. Pretreatment of Lignocellulosic Materials for Efficient Bioethanol Production. *Adv. Biochem. Eng./ Biotechnol.* 2007, *108*, 41–65.

(14) Jorgensen, H.; Kristensen, J.B. and Felby, C. Enzymatic Conversion of Lignocellulose into Fermentable Sugars: Challenges and Opportunities. Biofuels, *Bioprod. Bioref.* 2007, *1*, 119–134.

(15) Camassola, M. and Dillon, A.J.P. Biological Pretreatment of Sugar Cane Bagasse for the Production of Cellulases and Xylanases by Penicillium echinulatum. *Ind. Crops. Prod.* 2009, *29*, 642–647.

(16) Pandey, A.; Soccol, C.R.; Nigam, P.; Soccol, V.T.; Vandenberghe, L.P.S. and Mohan, R. Biotechnological Potential of Agro-Industrial Residues. II: Cassava Bagasse. *Bioresource Technol.* 2000, *74*, 81–87.

(17) Sun, Y. and Cheng, J. Hydrolysis of Lignocellulosic Materials for Ethanol Production: A Review. *Bioresource Technol.* 2002, *83*, 1–11.

(18) Perez, J.; Dorado, J.M.; Rubia, T.D. and Martinez, J. Biodegradation and Biological Treatment of Cellulose, Hemicellulose and Lignin: An Overview. *Int. Microbiol.* 2002, *5*, 53–63.

(19) Kumar, P.; Barrett, D.M.; Delwiche, M.J. and Stroeve, P. Methods for Pretreatment of Lignocellulosic Biomass for Efficient Hydrolysis and Biofuel Production. *Ind. Eng. Chem. Res.* 2009, *48*, 3713–3729.

(20) Kuhad, R.C.; Singh, A. and Eriksson, K.E. Microorganisms and Enzymes Involved in Degradation of Plant Fiber Cell Walls. *Adv. Biochem. Eng./Biotechnol.* 1997, *57*, 45–125.

(21) McMillan, J.D. Pretreatment of lignocellulosic biomass. In: Enzymatic Conversion of Biomass for Fuels Production. In: Himmel, Michael E., Baker, John O., Overend, Ralph P. (Ed.), American Chemical Society, Washington, DC, 1994, 292–324.

(22) Morjanoff, P.J. and Gray, P.P. Optimization of Steam Explosion as Method for Increasing Susceptibility of Sugarcane Bagasse to Enzymatic Saccharification. *Biotechnol. Bioeng.* 1987, *29*, 733–741.

(23) Payne, C.M.; Knott, B.C.; Mayes, H.B.; Hansson, H.; Himmel, M.E.; Sandgren, M.; Stahlberg, J. and Beckham, G.T. Fungal Cellulases. *Chem. Rev.* 2015, *115*, 1308–1448.

(24) Silverstein, R.A.; Chen, Y.; Sharma-Shivappa, R.R.; Boyette, M.D. and Osborne, J. A Comparison of Chemical Pretreatment

Methods for Improving Saccharification of Cotton Stalks. *Bioresource Technol.* 2007, *98*, 3000–3011.

(25) Millet, M.A.; Baker, A.J. and Scatter, L.D. Physical and Chemical Pretreatment for Enhancing Cellulose Saccharification. *Biotech. Bioeng. Symp.* 1976, *6*, 125–153.

(26) Cadoche, L. and Lopez, G.D. Assessment of Size Reduction as a Preliminary Step in the Production of Ethanol from Lignocellulosic Wastes. *Biol. Wastes.* 1989, *30*, 153–157.

(27) Zhu, J.Y.; Pan, X.; Ronald, S. and Zalesny, J. Pretreatment of Woody Biomass for Biofuel Production: Energy Efficiency, Technologies, and Recalcitrance. *Appl. Microbiol. Biotechnol.* 2010, *87*, 847–857.

(28) Barakat, A.; Laigle, C.M.; Solhy, A.; Arancon, R.A.D.; Vriesa, H. and Luquec, R. Mechanical Pretreatments of Lignocellulosic Biomass: Towards Facile and Environmentally Sound Technologies for Biofuels Production. *RSC Adv.*, 2014, *4*, 48109–48127.

(29) Madison, M.J.; Kelly, G.C.; Liang, C.; Karim, M.N.; Falls, M. and Holtzapple, M.T. Mechanical Pretreatment of Biomass e Part I: Acoustic and Hydrodynamic Cavitation. *Biomass Bioenerg.* 2017, *98*, 135–141.

(30) Kilzer, F.J.; Broido, A. Speculations on the Nature of Cellulose Pyrolysis. *Pyrodynamics* 1965, *2*, 151–163.

(31) Shafizadeh, F. and Bradbury, A.G.W. Thermal Degradation of Cellulose in Air and Nitrogen at Low Temperatures. *J. Appl. Polym. Sci.* 1979, *23*, 1431–1442.

(32) Fan, L.T.; Gharpuray, M.M.; Lee, Y.H. *Cellulose Hydrolysis*; Biotechnology Monographs; Springer: Berlin; *Vol. 3*, p 57.

(33) Zwart, R.W.R.; Boerrigter, H. and Van der Drift, A. The Impact of Biomass Pretreatment on the Feasibility of Overseas Biomass Conversion to Fischer-Tropsch Products. *Energ. Fuel.* 2006, *20*, 2192–2197.

(34) Undri, A.; Zaid, M.A.; Briens, C.; Berruti, F.; Rosi, L.; Bartoli, M.; Frediani, M. and Frediani, P. Bio-oil from Pyrolysis of Wood Pellets using a Microwave Multimode Oven and Different Microwave Absorbers. *Fuel.* 2015, *153*, 464–482.

(35) Bartoli, M.; Rosi, L.; Giovannelli, A.; Frediani, P. and Frediani, M. Bio-oil from residues of short rotation coppice of poplar using amicrowave assisted pyrolysis. *J. Anal. Appl. Pyrol.*, 2016, *119*, 224–232.

(36) Heo, H.S.; Park, H.J.; Park, Y.K.; Ryu, C.; Suh, D.J.; Suh, Y.W.; Yim, J.H. and Kim, S.S. Bio-Oil Production from Fast Pyrolysis of Waste Furniture Sawdust in a Fluidized Bed. *Bioresource Technol.* 2010, *101*, S91–S96.

(37) Asadullah, M.; Rahman, M.A.; Ali, M.M.; Motin, M.A.; Sultan, M.B.; Alam, M.R. and Rahman, M.S. Jute Stick Pyrolysis for Bio-Oil Production in Fluidized Bed Reactor. *Bioresource Technol.* 2008, *99*, 44–50.

(38) Boateng, A.A.; Daugaard, D.E.; Goldberg, N.M. and Hicks, K.B. Bench Scale Fluidized Bed Pyrolysis of Switchgrass for Bio-Oil Production. *Ind. Eng. Chem. Res.* 2007, *46*, 1891–1897.

(39) Wang, H.; Male, J. and Wang, Y. Recent Advances in Hydrotreating of Pyrolysis Bio-Oil and Its Oxygen-Containing Model Compounds. *ACS Catal.* 2013, *3*, 1047–1070.

(40) Czernik, S. and Bridgwater, A. Overview of Applications of Biomass Fast Pyrolysis Oil. *Energ. Fuels.* 2004, *18*, 590–598.

(41) Yildiz, G.; Pronk, M.; Djokic, M.; van Geem, K.M.; Ronsse, F.; Van Duren, R. and Prins, W. Validation of a New Set-up for Continuous Catalytic Fast Pyrolysis of Biomass Coupled with Vapour Phase Upgrading. *J. Anal. Appl. Pyrol.* 2013, *103*, 343–351.

(42) Ruddy, D.A.; Schaidle, J.A.; Ferrell, J.R.; Wang, J.; Moens, L. and Hensley, J.E. Recent Advances in Heterogeneous Catalysts for Bio-oil Upgrading via Ex Situ Catalytic Fast Pyrolysis: Catalyst Development through the Study of Model Compounds. *Green Chem.* 2014, *16*, 454–490.

(43) Liu, C.; Wang, H.; Karim, A.M.; Sun, J. and Wang, Y. Catalytic Fast Pyrolysis of Lignocellulosic Biomass. *Chem. Soc. Rev.* 2014, *43*, 7594–7623.

(44) Wan, S. and Wang, Y. A Review on Ex Situ Catalytic Fast Pyrolysis of Biomass. *Front. Chem. Sci. Eng.* 2014, *8*, 280–294.

(45) Zhang, H.; Xiao, R.; Huang, H. and Xiao, G. Comparison of Non-Catalytic and Catalytic Fast Pyrolysis of Corncob in a Fluidized Bed Reactor. *Bioresource Technol.* 2009, *100*, 1428–1434.

(46) Mukarakate, C.; Zhang, X.; Stanton, A.R.; Robichaud, D.J.; Ciesielski, P.N.; Malhotra, K.; Donohoe, B.S.; Gjersing, E.; Evans, R.J.; Heroux, D.S.; Richards, R.; Lisa, K. and Nimlos, M.R. Real Time Monitoring of the Deactivation of HZSM-5 During Upgrading of Pine Pyrolysis Vapors. *Green Chem.* 2014, *16*, 1444–1461.

(47) Adjaye, J.D. and Bakhshi, N. Production of Hydrocarbons by Catalytic Upgrading of a Fast Pyrolysis Bio-Oil. Part I: Conversion over Various Catalysts. *Fuel Process. Technol.* 1995, *45*, 161–183.

(48) Jae, J.; Tompsett, G.A.; Foster, A.J.; Hammond, K.D.; Auerbach, S.M.; Lobo, R.F. and Huber, G.W. Investigation into the Shape Selectivity of Zeolite Catalysts for Biomass Conversion. *J. Catal.* 2011, *279*, 257–268.

(49) Mihalcik, D.J.; Mullen, C.A. and Boateng, A.A. Screening Acidic Zeolites for Catalytic Fast Pyrolysis of Biomass and Its Components. *J. Anal. Appl.Pyrol.* 2011, *92*, 224–232.

(50) Jackson, M.A.; Compton, D.L. and Boateng, A.A. Screening Heterogeneous Catalysts for the Pyrolysis of Lignin. *J. Anal. Appl. Pyrol.* 2009, *85*, 226–230.

(51) Mukarakate, C.; Watson, M.J.; ten Dam, J.; Baucherel, X.; Budhi, S.; Yung, M.M.; Ben, H.; Iisa, K.; Baldwin, R.M. and Nimlos, M.R. Upgrading Biomass Pyrolysis Vapors over β-Zeolites: Role of Silica-to-Alumina Ratio. *Green Chem.* 2014, *16*, 4891–4905.

(52) Bridgwater, A.V. Review of Fast Pyrolysis of Biomass and Product Upgrading. *Biomass Bioenerg.* 2012, *38*, 68–94.

(53) Zhang, H.; Cheng, Y.T.; Vispute, T.P.; Xiao, R. and Huber, G.W. Catalytic Conversion of Biomass Derived Feed stocks into Olefins and Aromatics with ZSM-5: The Hydrogen to Carbon Effective Ratio. *Energ. Environ. Sci.* 2011, *4*, 2297–2307.

(54) Scahill, J.; Diebold, J. and Porwer, A. Engineering Aspects of Upgrading Pyrolysis Oil Using Zeolites. Chapter: Research in Thermochemical Biomass Conversion. Springer; 1988, pp. 927–940.

(55) Carlson, T.R.; Jae, J. and Huber, G.W. Mechanistic Insights from Isotopic Studies of Glucose Conversion to Aromatics over ZSM-5. *Chem. Cat. Chem.* 2009, *1*, 107–110.

(56) Cheng, Y.T.; Jae, J.; Shi, J.; Fan, W. and Huber, G.W. Production of Renewable Aromatic Compounds by Catalytic Fast Pyrolysis of Lignocellulosic Biomass with Bifunctional Ga/ZSM-5 Catalysts. *Angew. Chem.-Ger Edit.* 2012, *124*, 1416–1419.

(57) Horne, P.A. and Williams, P.T. The Effect of Zeolite ZSM-5 Catalyst Deactivation During the Upgrading of Biomass-Derived Pyrolysis Vapors. *J. Anal. Appl. Pyrol.* 1995, *34*, 65–85.

(58) Prasomsri, T.; Nimmanwudipong, T. and Román-Leshkov, Y. Effective Hydrodeoxygenation of Biomass-Derived Oxygenates into Unsaturated hydrocarbons by MoO_3 using low H_2 pressures. *Energy Environ. Sci.* 2013, *6*, 1732–1738.

(59) Prasomsri, T.; Shetty, M.; Murugappan, K. and Román-Leshkov, Y. Insights into the Catalytic Activity and Surface Modification of MoO_3 During the Hydrodeoxygenation of Lignin-Derived Model Compounds into Aromatic Hydrocarbons Under Low Hydrogen Pressures. *Energ. Environ. Sci.* 2014, *7*, 2660–2669.

(60) Shetty, M.; Murugappan, K.; Prasomsri, T.; Green, W.H. and Román-Leshkov,Y. Reactivity and Stability Investigation of Supported Molybdenum Oxide Catalysts for the Hydrodeoxygenation (HDO) of m-Cresol. *J. Catal.* 2015, *331*, 86–97.

(61) Lee, W.S.; Wang, Z.S.; Wu, R.J. and Bhan, A. Selective Vapor-Phase Hydrodeoxygenation of Anisole to Benzene on Molybdenum Carbide Catalysts. *J. Catal.* 2014, *319*, 44–53.

(62) Sullivan, M.M. and Bhan, A. Acetone Hydrodeoxygenation over Bifunctional Metallic-Acidic Molybdenum Carbide Catalysts. *ACS Catal.* 2016, *6*, 1145–1152.

(63) Lee, W.-S.; Kumar, A.; Wang, Z. and Bhan, A. Chemical Titration and Transient Kinetic Studies of Site Requirements in Mo_2C-Catalyzed Vapor Phase Anisole Hydrodeoxygenation. *ACS Catal.* 2015, *5*, 4104–4114.

(64) Sullivan, M.M.; Held, J.T. and Bhan, A. Structure and Site Evolution of Molybdenum Carbide Catalysts upon Exposure to Oxygen. *J. Catal.* 2015, *326*, 82–91.

(65) Budhi, S.; Mukarakate, C.; Iisa, K.; Pylypenko, S.; Ciesielski, P.N.; Yung, M.M.; Donohoe, B.S.; Katahira, R.; Nimlos, M.R. and Trewyn, B.G. Molybdenum Incorporated Mesoporous Silica Catalyst for Production of Biofuels and Value-Added Chemicals via Catalytic Fast Pyrolysis. *Green Chem.* 2015, *17*, 3035–3046.

(66) Nolte, M.W.; Zhang, J. and Shanks, B.H. Ex Situ Hydrodeoxygenation in Biomass Pyrolysis using Molybdenum Oxide and Low Pressure Hydrogen. *Green Chem.* 2016, *18*, 134–138.

(67) Murugappan, K.; Mukarakate, C.; Budhi, S.; Shetty, M.; Nimlos, M.R. and Román-Leshkov, Y. Supported Molybdenum Oxides as Effective Catalysts for the Catalytic Fast Pyrolysis of Lignocellulosic Biomass. *Grren Chem.* 2016, *18*, 5548–5557.

(68) Kumakura, M. and Kaetsu, I. Effect of Radiation Pretreatment of Bagasse on Enzymatic and Acid Hydrolysis. *Biomass.* 1983, *3*, 199–208.

(69) Jin, S.B.; Ja, K.K.; Young, H.H.; Byung, C.L.; In-Geol, C. and Heon, K.K. Improved Enzymatic Hydrolysis Yield of Rice Straw Using Electron Beam Irradiation Pretreatment. *Bioresour. Technol.* 2009, *100*, 1285–1290.

(70) Takacs, E.; Wojnarovits, L.; Foldvary, C.; Hargittai, P.; Borsa, J. and Sajo, I. Effect of Combined Gamma-Irradiation and Alkali Treatment on Cotton Cellulose. *Radiat. Phys. Chem.* 2000, *57*, 399–403.

(71) Bak, J.S. Electron Beam Irradiation Enhances the Digestibility and Fermentation Yield of Water-Soaked Lignocellulosic Biomass. *Biotechnol. Reports.* 2014, *4*, 30–33.

(72) Karthika, K.I.; Arun, A.B. and Rekha, P.D. Enzymatic Hydrolysis and Characterization of Lignocellulosic Biomass Exposed to Electron Beam Irradiation. *Carbohydr Polym.* 2012, *90*, 1038–1045.

(73) Loow, Y.L.; Wu, T.Y.; Yang, G.H.; Jahim, J.M.; Teoh, W.H. and Mohammad, A.W. Role of Energy Irradiation in Aiding Pretreatment of Lignocellulosic Biomass for Improving Reducing Sugar Recovery, *Cellulose*. 2016, *23*, 2761–2789.

(74) Xiong, J.; Ye, J.; Liang, W.Z. and Fan, P.M. Influence of Microwave on the Ultrastructure of Cellulose I. *J. South China Univ. Technol.* 2000, *28*, 84–89.

(75) Azuma, J.; Tanaka, F. and Koshijima, T. Enhancement of Enzymatic Susceptibility of Lignocellulosic Wastes by Microwave Irradiation. *J. Ferment. Technol.* 1984, *62*, 377–384.

(76) Ooshima, H.; Aso, K. and Harano, Y. Microwave Treatment of Cellulosic Materials for Their Enzymatic Hydrolysis. *Biotechnol. Lett.* 1984, *6*, 289–294.

(77) Intanakul, P.; Krairish, M. and Kitchaiya, P. Enhancement of Enzymatic Hydrolysis of Lignocellulosic Wastes by Microwave Pretreatment under Atmospheric Pressure. *J. Wood Chem. Technol.* 2003, *23*, 217–225.

(78) Zhu, S.; Wu, Y.; Yu, Z.; Wang, C.; Yu, F.; Jin, S.; Ding, Y.; Chi, R.; Liao, J. and Zhang, Y. Comparison of Three Microwave/Chemical Pretreatment Processes for Enzymatic Hydrolysis of Rice Straw. *Biosyst. Eng.* 2005, *93*, 279–283.

(79) Richel, A. and Jacquet, N. Microwave-assisted Thermochemical and Primary Hydrolytic Conversions of Lignocellulosic Resources: A Review. *Biomass Conv. Bioref.* 2015, *5*, 115–124.

(80) Sun, R.C. and Tomkinson, R.C. Characterization of Hemicelluloses Obtained by Classical and Ultrasonically Assisted Extractions from Wheat Straw. *Carbohyd. Polym.* 2002, *50*, 263–271.

(81) Yachmenev, V.; Condon, B.; Klasson, T. and Lambert, A. Acceleration of the Enzymatic Hydrolysis of Corn Stover and Sugarcane Bagasse Celluloses by Low Intensity Uniform Ultrasound. *J. Biobased Mater. Bio.* 2009, *3*, 25–31.

(82) Karimia, M.; Jenkinsb, B. and Stroevea, P. Ultrasound Irradiation in the Production of Ethanol from Biomass, *Renew. Sust. Energ. Rev.* 2014, *40*, 400–421.

(83) Luo, J.; Fang, Z. and Smith, R.L. Ultrasound-Enhanced Conversion of Biomass to Biofuels. *Prog. Energ. Combust.* 2014, *41*, 56–93.

(84) Chang, D.C.; Chassy, B.M.; Saunders, J.A. and Sowers, A.E. Guide to Electroporation and Electrofusion; Academic Press: San Diego, Eds. 1, 1992.

(85) Ho, S.Y. and Mittal, G.S. Electroporation of Cell Membranes: A Review. *Crit. Rev. Biotechnol.* 1996, *16*, 349–362.

(86) Knorr, D. and Angersbach, A. Impact of High Intensity Electrical Field Pulses on Plant Membrane Permeabilization. *Trends Food Sci. Technol.* 1998, *9*, 185–191.

(87) Lynch, P.T.; Davey, M.R. Electrical Manipulation of Cells. Chapman and Hall: New York, 1996.

(88) Zimmermann, U.; Neil, G.A. Electromanipulation of Cells. CRC Press: Boca Raton, FL, 1996.

(89) Vangerbasch, A.; Heinz, V. and Knorr, D. Effects of Pulsed Electric Fields on Cell Membranes in Real Food Systems. *Innov. Food Sci. Emerg. Technol.* 2000, *1*, 135–149.

(90) Jayaram, S.; Catle, G.S.P. and Margaritis, A. The Effect of High Field DC Pulse and Liquid Medium Conductivity on Survivability of Lactobacillus Brevis. *Appl. Microbiol. Biotechnol.* 1993, *40*, 117–122.

(91) Giner, G.; Gimeno, V.; Barbosa-Canovas, G.V. and Martin, O. Effects of Pulsed Electric Field Processing on Apples and Pear Polyphenoloxidases. *Food Sci. Technol. Int.* 2001, *7*, 339–345.

(92) Taiwo, K.A.; Angerbasch, A.; Ade-Omowaye, B.I.O. and Knorr, D. Effects of Pretreatments on the Diffusion Kinetics and Some Quality Parameters of Osmotically Dehydrated Apple Slices. *J. Agric. Food Chem.* 2001, *49*, 2804–2811.

(93) Eshtiaghi, M.N. and Knorr, D. High Electric Field Pulse Treatment: Potential for Sugar Beet Processing. *J. Food Eng.* 2002, *52*, 265–272.

(94) Bazhal, M.I.; Lebvoka, N.I. and Vorobiev, E. Pulsed Electric Field Treatment of Apple Tissue during Compression for Juice Extraction. *J. Food Eng.* 2001, *50*, 129–139.

(95) Rastogi, N.K.; Eshtiaghi, M.N. and Knorr, D. Accelerated Mass Transfer during Osmotic Dehydration of High Intensity Electric Field Pulse Treated Potatoes. *J. Food Sci.* 1999, *64*, 1020–1023.

(96) Kumar, P.; Barrett, D.M.; Delwiche, M.J. and Stroeve, P. Pulsed Electric Field Pretreatment of Switch Grass and Woodchips Species for Biofuels Production. *Ind. Eng. Chem. Res.* 2011, *50*, 10996–11001.

(97) Kumar, P.; Barrett, M.D.; Delwiche, J.M. and Stroeve, P. Methods for Pretreatment of Lignocellulosic Biomass for Efficient Hydrolysis and Biofuel Production. *Ind. Ind. Eng. Chem. Res.* 2009, *48*, 3713–3729.

(98) Zbinden, M.D.A.; Sturm, B.S.M.; Nord, R.D.; Carey, W.J.; Moore, D.; Shinogle, H. and Stagg-Williams, S.M. Pulsed Electric Field (PEF) as an Intensification Pretreatment for Greener Solvent Lipid Extraction from Microalgae. *Biotechnol. Bioeng.* 2013, *110*, 1605–1615.

(99) Yu, X.; Gouyo, T.; Grimi, N.; Bals, O. and Vorobiev, E. Pulsed Electric Field Pretreatment of Rapeseed Green Biomass (Stems) to Enhance Pressing and Extractives Recovery. *Bioresource Technol.* 2016, *199*, 194–201.

(100) Kabel, M.A.; Bos, G.; Zeevalking, J.; Voragen, A.G. and Schols, H.A. Effect of Pretreatment Severity on Xylan Solubility and Enzymatic Breakdown of the Remaining Cellulose from Wheat Straw. *Bioresource Technol.* 2007, *98*, 2034–2042.

(101) Li, J.; Henriksson, G. and Gellerstedt, G. Lignin Depolymerization/ Repolymerization and Its Critical Role for Delignification of Aspen Wood by Steam Explosion. *Bioresource Technol.* 2007, *98*, 3061–3068.

(102) Duff, S.J.B. and Murray, W.D. Bioconversion of Forest Products Industry Waste Cellulosics to Fuel Ethanol: A Review. *Bioresource Technol.* 1996, *55*, 1–33.

(103) Weil, J.R.; Sarikaya, A.; Rau, S.L.; Goetz, J.; Ladisch, C.M.; Brewer, M.; Hendrickson, R. and Ladisch, M.R. Pretreatment of Yellow Poplar Sawdust by Pressure Cooking in Water. *Appl. Biochem. Biotechnol.* 1997, *68*, 21–40.

(104) Baugh, K.D.; Levy, J.A. and McCarty, P.L. Thermochemical Pretreatment of Lignocellulose to Enhance Methane Fermentation: I. Monosaccharide and Furfurals Hydrothermal Decomposition and Product Formation Rates. *Biotechnol. Bioeng.* 1988, *31*, 50–61.

(105) Ballesteros, I.; Negro, M.J.; Oliva, J.M.; Cabanas, A.; Manzanares, P. and Ballesteros, M. Ethanol Production from Steam Explosion Pretreated Wheat Straw. *Appl. Biochem. Biotechnol.* 2006, *70*, 3–15.

(106) Stenberg, K.; Tengborg, C.; Galbe, M. and Zacchi, G. Optimization of Steam Pretreatment of SO_2 Impregnated Mixed Softwoods for Ethanol Production. *J. Chem. Technol. Biotechnol.* 1998, *71*, 299–308.

(107) Wright, J.D. Ethanol from Biomass by Enzymatic Hydrolysis. *Chem. Eng. Prog.* 1998, *84*, 62–74.

(108) Holtzapple, M.T.; Humphrey, A.E. and Taylor, J.D. Energy Requirements for the Size Reduction of Poplar and Aspen Wood. *Biotechnol. Bioeng.* 1989, *33*, 207–210.

(109) Kobayashi, F.; Take, H.; Asada, C. and Nakamura, Y. Methane Production from Steam-Exploded Bamboo. *J. Biosci. Bioeng.* 2004, *97*, 426–428.

(110) Ballesteros, M.J.; Oliva, I.; Negro, M.J.; Manzanares, P. and Ballesteros, M. Enzymic Hydrolysis of Steam Exploded Herbaceous Agricultural Waste (*Brassica Carinata*) at Different Particle Sizes. *Process Biochem.* 2002, *38*, 187–192.

(111) Viola, E.; Zimabardi, F.; Cardinale, M.; Cardinale, G.; Braccio, G. and Gamabacorta, E. Processing Cereal Straws by Steam Explosion in a Pilot Plant to Enhance Digestibility in Ruminants. *Bioresource Technol.* 2008, *99*, 681–689.

(112) Cara, C.; Ruiz, C.; Ballesteros, M.; Manzanares, P.; Negro, M.J. and Castro, E. Production of Fuel Ethanol from Steam-Explosion Pretreated Olive Tree Pruning. *Fuel*. 2008, *87*, 692–700.

(113) Mackie, K.L.; Brownell, H.H.; West, K.L. and Saddler, J.N. Effect of Sulphur Dioxide and Sulphuric Acid on Steam Explosion of Aspen Wood. *J. Wood Chem. Technol.* 1985, *5*, 405–425.

(114) Bobleter, O.; Binder, H.; Concin, R.; Burtscher, E. The Conversion of Biomass to Fuel Raw Material by Hydrothermal Pretreatment. In: Palz, W., Chartier, P., Hall, D.O. (Eds.), Energy from Biomass. Applied Science Publishers: London, 1981, 554–562.

(115) Mok, W.S.L. and Antal, M.J. Jr. Uncatalyzed Solvolysis of Whole Biomass Hemicellulose by Hot Compressed Liquid Water. *Ind. Eng. Chem. Res.* 1992, *31*, 1157–1161.

(116) Bouchard, J.; Nguyen, T.S.; Chornet, E. and Overend, R.P. Analytical Methodology for Biomass Pretreatment. Part 2: Characterization of the Filtrates and Cumulative Product Distribution as a Function of Treatment Severity. *Bioresour. Technol.* 1991, *36*, 121–131.

(117) Allen, S.G.; Kam, L.C.; Zemann, A.J. and Antal, M.J. Fractionation of Sugar Cane with Hot, Compressed, Liquid Water. *Ind. Eng. Chem. Res.* 1996, *35*, 2709–2715.

(118) Weil, J.R.; Sarikaya, A.; Rau, S.-L.; Goetz, J.; Ladisch, C.M.; Brewer, M.; Hendrickson, R. and Ladisch, M.R. Pretreatment of Yellow Poplar Sawdust by Pressure Cooking in Water. *Appl. Biochem. Biotechnol.* 1997, *681*, 21–40.

(119) Cara, C.; Moya, M.; Ballesteros, I.; Negro, M.J.; Gonzalez, A. and Ruiz, E. Influence of Solid Loading on Enzymatic Hydrolysis of Steam Exploded or Liquid Hot Water Pretreated Olive Tree Biomass. *Process Biochem.* 2007, *42*, 1003–1009.

(120) Perez, J.A.; Gonzalez, A.; Oliva, J.M.; Ballesteros, I. and Manzanares, P. Effect of Process Variables on Liquid Hot Water Pretreatment of Wheat Straw for Bioconversion to Fuel-Ethanol in a Batch Reactor. *J. Chem. Technol. Biotechnol.* 2007, *82*, 929–938.

(121) Tabata, T.; Yoshiba, Y.; Takashina, T.; Hieda, K. and Shimizu, N. Bioethanol Production from Steam-Exploded Rice Husk by Recombinant Escherichia Coli KO11. *World J. Microbiology Biotechnol.* 2017, *33*, 47.

(122) Auxenfans, T.; Crônier, D.; Chabbert, B. and Paes, G. Understanding the Structural and Chemical Changes of Plant Biomass Following Steam Explosion Pretreatment. *Biotechnol. Biofuels.* 2017, *10*, 36.

(123) Wood, I.P.; Cao, H.-G.; Tran, L.; Cook, N.; Ryden, P.; Wilson, D.R.; Moates, G.K.; Collins, S.R.A.; Elliston, A. and Waldron, K.W. Comparison of Saccharification and Fermentation of Steam Exploded Rice Straw and Rice Husk. *Biotechnol. Biofuels.* 2016, *9*, 193.

(124) Monschein, M. and Nidetzky, B. Effect of Pretreatment Severity in Continuous Steam Explosion on Enzymatic Conversion of Wheat Straw: Evidence from Kinetic Analysis of Hydrolysis Time Courses. *Bioresource Technol.* 2016, *200*, 287–296.

(125) Ferro, M.D.; Fernandes, M.C.; Paulino, A.F.C.; Prozil, S.O.; Gravitis, J.; Evtuguin, D.V. and Xavier, A.M.R.B. Bioethanol Production from Steam Explosion Pretreated and Alkali Extracted Cistus Ladanifer (Rockrose). *Biochem. Eng. J.* 2015, *104*, 98–105.

(126) Bauer, A.; Lizasoain, J.; Theuretzbacher, F.; Agger, J.W.; Rincón, M.; Menardo, S.; Saylor, M.K.; Enguídanos, R.; Nielsen, P.J.; Potthast, A.; Zweckmair, T.; Gronauer, A. and Horn, S.J. Steam Explosion Pretreatment for Enhancing Biogas Production of Late Harvested Hay. *Bioresource Technol.* 2014, *166*, 403–410.

(127) Iroba, K.L.; Tabil, L.G.; Sokhansanj, S. and Dumonceaux, T. Pretreatment and Fractionation of Barley Straw Using Steam Explosion at Low Severity Factor. *Biomass Bioenerg.* 2014, *66*, 286–300.

(128) Alizadeh, H.; Teymouri, F.; Gilbert, T.I. and Dale, B.E. Pretreatment of Switch Grass by Ammonia Fiber Explosion (AFEX). *Appl. Biochem. Biotechnol.* 2005, *121*, 1133–1141.

(129) Mes-Hartree, M.; Dale, B.E. and Craig, W.K. Comparison of Steam and Ammonia Pretreatment for Enzymatic Hydrolysis of Cellulose. *Appl. Microbic ol. Biotechnol.* 1988, *29*, 462–468.

(130) Gollapalli, L.E.; Dale, B.E. and Rivers, D.M. Predicting Digestibility of Ammonia Fiber Explosion (AFEX) Treated Rice Straw. *Appl. Biochem. Biotechnol.* 2002, *98*, 23–35.

(131) Holtzapple, M.T.; Jun, J.H.; Ashok, G.; Patibandla, S.L. and Dale, B.E. The Ammonia Freeze Explosion (AFEX) Process: A Practical Lignocellulose Pretreatment. *Appl. Biochem. Biotechnol.* 1991, *28*, 59–74.

(132) Teymouri, F.; Perez, L.L.; Alizadeh, H. and Dale, B.E. Ammonia Fiber Explosion Treatment of Corn Stover. *Appl. Biochem. Biotechnol.* 2004, *113–116*, 951–963.

(133) Murnen, H.K.; Balan, V.; Chundawat, S.P.S.; Bals, B.; Sousa, L.D.C. and Dale, B.E. Optimization of Ammonia Fiber Expansion (AFEX) Pretreatment and Enzymatic Hydrolysis of Miscanthus x Giganteus to Fermentable Sugars. *Biotechnol. Prog.* 2007, *23*, 846–850.

(134) Isci, A.; Himmelsbach, J.N.; Pometto, A.L.; Raman, R. and Anex, R.P. Aqueous Ammonia Soaking of Switch Grass Followed by

Simultaneous Saccharification and Fermentation. *Appl. Biochem. Biotechnol.* 2008, *144*, 69–77.

(135) Kim, T.H. and Lee, Y.Y. Pretreatment of Corn Stover by Soaking in Aqueous Ammonia. *Appl. Biochem. Biotechnol.* 2005, *124*, 1119–1132.

(136) Kim, T.H. and Lee, Y.Y. Pretreatment and Fractionation of Corn Stover by Ammonia Recycle Percolation Process. *Bioresource Technol.* 2005, *96*, 2007–2013.

(137) Mosier, N.; Wyman, C.; Dale, B.; Elander, R.; Holtzapple, Y.Y.L.M. and Ladisch, M. Features of Promising Technologies for Pretreatment of Lignocellulosic Biomass. *Bioresource Technol.* 2005, *96*, 673–686.

(138) Lee Y.J. Oxidation of Sugarcane Bagasse Using a Combination of Hypochlorite and Peroxide. Master's Thesis, Department of Food Science, Graduate Faculty of the Louisiana State University and Agricultural and Mechanical College, 2005.

(139) Kamm, B.; Leib, S.B.; Schönicke, P. and Bierbaum, M. Biorefining of Lignocellulosic Feedstock by a Modified Ammonia Fiber Expansion Pretreatment and Enzymatic Hydrolysis for Production of Fermentable Sugars. *Chem. Sus. Chem.* 2017, *10*, 48–52.

(140) Ong, R.G.; Higbee, A.; Bottom, S.; Dickinson, Q.; Xie, D.; Smith, S.A.; Serate, J.; Pohlmann, E.; Jones, A.D.; Coon, J.J.; Sato, T.K.; Sanford, G.R.; Eilert, D.; Oates, L.G.; Piotrowski, J.S.; Bates, D.M.; Cavalier, D. and Zhang, Y. Inhibition of Microbial Biofuel Production in Drought-Stressed Switchgrass Hydrolysate. *Biotechnol. Biofuels.* 2016, *9*, 1–14.

(141) Lee, W.-C. and Kuan, W.-C. Miscanthus as Cellulosic Biomass for Bioethanol Production. *Biotechnol J.* 2015, *10*, 840–854.

(142) Zhao, C.; Ding, W.; Chen, F.; Cheng, C. and Shao, Q. Effects of Compositional Changes of AFEX-Treated and H-AFEX-Treated Corn Stover on Enzymatic Digestibility. *Bioresource Technol.* 2014, *155*, 34–40.

(143) Dale, B.E. and Moreira, M.J. A Freeze-Explosion Technique for Increasing Cellulose Hydrolysis. *Biotechnol. Bioeng. Symp.* 1982, *12*, 31–43.

(144) Zheng, Y.Z.; Lin, H.M. and Tsao, G.T. Pretreatment for Cellulose Hydrolysis by Carbon Dioxide Explosion. *Biotechnol. Prog.* 1998, *14*, 890–896.

(145) Narayanaswamy, N.; Faik, A.; Goetz, J.D. and Gu, T. Supercritical Carbon Dioxide Pretreatment of Corn Stover and Switchgrass for Lignocellulosic Ethanol Production. *Bioresource Technol.* 2011, *102*, 6995–7000.

(146) Srinivasan, N. and Ju, L.-K. Statistical Optimization of Operating Conditions for Supercritical Carbon Dioxide-Based Pretreatment of Guayule Bagasse. *Biomass Bioenerg.* 2012, *47*, 451–458.

(147) Gu, T.; Held, M.A. and Faik, A. Supercritical CO_2 and Ionic Liquids for the Pretreatment of Lignocellulosic Biomass in Bioethanol Production. *Environ. Technol.* 2013, *34*, 1735–1749.

(148) Capolupo, L. and Faraco, V. Green Methods of Lignocellulose Pretreatment for Biorefinery Development. *Appl. Microbiol. Biot.* 2016, *100*, 9451–9467.

(149) Carneiro, T.F.; Timko, M.; Prado, J.M. and Berni, M. Biomass Fractionation Technologies for a Lignocellulosic Feedstock Based Biorefinery. *Book Chapter: Biomass Pretreatment with Carbon Dioxide*, Book Chapter 1, 2016, pp. 385–407.

(150) Palonen, H.; Thomsen, A.B.; Tenkanen, M.; Schmidt, A.S. and Viikari, L. Evaluation of Wet Oxidation Pretreatment for Enzymatic Hydrolysis of Softwood. *Appl. Biochem. Biotechnol.* 2004, *117*, 1–17.

(151) Olsson, L.; Jörgensen, H.; Krogh, K.B.R.; Roca, C. Bioethanol Production from Lignocellulosic Material. Polysaccharides Structural Diversity and Functional Versatility. In: Dimitriu, S. (Ed.), Marcel Dekker: New York, 2005, 957–993.

(152) Martín, C.; Thomsen, M.H.; Hauggaard, H. and Thomsem, A.B. Wet Oxidation Pretreatment, Enzymatic Hydrolysis and Simultaneous Saccharification and Fermentation of Clover-Ryegrass Mixtures. *Bioresource Technol.* 2008, *99*, 8777–8782.

(153) Klinke, H.B.; Ahring, B.K.; Schmidt, A.S. and Thomsen, A.B. Characterization of Degradation Products from Alkaline Wet oxidation of Wheat Straw. *Bioresource Technol.* 2002, *82*, 15–26.

(154) Ayeni, A.O.; Banerjee, S.; Omoleye, J.A.; Hymore, F.K.; Giri, B.S.; Deshmukh, S.C.; Pandey, R.A. and Mudliar, S.N. Optimization of Pretreatment Conditions Using Full Factorial Design and Enzymatic Convertibility of shea tree sawdust. *Biomass Bioenerg.* 2013, *48*, 130–138.

(155) Fang, G.-G.; Liu, S.-S. and Shen, K.-Z. Pretreatment of Poplar Wood Residues Using Wet Oxidation to Enhance Enzymatic Digestibility. *Chung-kuo Tsao Chih/China Pulp Paper.* 2015, *34*, 6–12.

(156) Srinivas, K.; Oliveira, F.C.; Teller, P.J.; Gonçalves, A.R.; Helms, G.L. and Ahring, B.K. Oxidative Degradation of Biorefinery Lignin Obtained After Pretreatment of Forest Residues of Douglas Fir. *Bioresource Technol.* 2016, *221*, 394–404.

(157) Katsimpouras, C.; Kalogiannis, K.G.; Kalogianni, A.; Lappas, A.A. and Topakas, E. Production of High Concentrated Cellulosic

Ethanol by Acetone/Water Oxidized Pretreated Beech Wood. *Biotechnol. Biofuels.* 2017, *10*, 54.

(158) Sun, Y. and Cheng, J. Hydrolysis of Lignocellulosic Materials for Ethanol Production: A Review. *Bioresource Technol.* 2002, *83*, 1–11.

(159) Ben-Ghedalia, D. and Miron, J. The Effect of Combined Chemical and Enzyme Treatment on the Saccharification and in Vitro Digestion Rate of Wheat Straw. *Biotechnol. Bioeng.* 1981, *23*, 823–831.

(160) Neely, W.C. Factors Affecting the Pretreatment of Biomass with Gaseous Ozone. *Biotechnol. Bioeng.* 1984, *26*, 59–65.

(161) Ben-Ghedalia, D. and Shefet, G. Chemical Treatments for Increasing the Digestibility of Cotton Straw. *J. Agric. Sci.* 1983, *100*, 393–400.

(162) Vidal, P.F. and Molinier, J. Ozonolysis of Lignins Improvement of In Vitro Digestibility of Poplar Sawdust. *Biomass.* 1988, *16*, 1–17.

(163) Quesada, J.; Rubio, M. and Gomez, D. Ozonation of Lignin Rich Solid Fractions from Corn Stalks. *J. Wood Chem. Technol.* 1999, *19*, 115–137.

(164) Lasry, T.; Laurent, J.L.; Euphrosine-Moy, V.; Bes, R.S.; Molinier, J. and Mathieu, I. Identification and Evaluation of Polar sawdust Ozonation Products. *Analysis.* 1990, *18*, 192–199.

(165) Euphrosine-Moy, V.; Lasry, T.; Bes, R.S.; Molinier, J. and Mathieu, J. Degradation of Poplar Lignin with Ozone. *Ozone Sci. Eng.* 1991, *13*, 239–248.

(166) Morrison, W.H. and Akin, D.E. Water Soluble Reaction Products from Ozonolysis of Grasses. *J. Agric. Food Chem.* 1990, *38*, 678–681.

(167) Panneerselvam, A.; Sharma-Shivappa, R.R.; Kolar, P.; Ranney, T. and Peretti, S. Potential of Ozonolysis as a Pretreatment for Energy Grasses. *Bioresource Technol.* 2013, *148*, 242–248.

(168) Souza-Corrêa, J.A.; Oliveira, C.; Wolf, L.D.; Nascimento, V.M.; Rocha, G.J.M. and Amorim, J. Atmospheric Pressure Plasma Pretreatment of Sugarcane Bagasse: The Influence of Moisture in the Ozonation Process. *Appl. Biochem. Biotech.* 2013, *171*, 104–116.

(169) Bellido, C.; Pinto, M.L.; Coca, M.; González-Benito, G. and García-Cubero, M.T. Acetone-Butanol-Ethanol (ABE) Production by Clostridium Beijerinckii from Wheat Straw Hydrolysates: Efficient Use of Penta and Hexa Carbohydrates. *Bioresource Technol.* 2014, *167*, 198–205.

(170) Bhattarai, S.; Bottenus, D.; Ivory, C.F.; Gao, A.H.; Bule, M.; Garcia-Perez, M. and Chen, S. Simulation of the Ozone

Pretreatment of Wheat Straw. *Bioresource Technol.* 2015, *196*, 78–87.

(171) Li, C.; Wang, L.; Chen, Z.; Li, Y.; Wang, R.; Luo, X.; Cai, G.; Li, Y.; Yu, Q. and Lu, J. Ozonolysis Pretreatment of Maize Stover: The Interactive Effect of Sample Particle Size and Moisture on Ozonolysis Process. *Bioresource Technol.* 2015, *183*, 240–247.

(172) Travaini, R.; Barrado, E. and Bolado, S. Effect of Ozonolysis Parameters on the Inhibitory Compound Generation and on the Production of Ethanol by Pichia Stipitis and Acetone-Butanol-Ethanol by Clostridium from Ozonated and Water Washed Sugarcane Bagasse. *Bioresource Technol.* 2016, *218*, 850–858.

(173) Sumphanwanich, J.; Leepipatpiboon, N.; Srinorakutara, T. and Akaracharanya, A. Evaluation of Dilute Acid Pretreated Bagasse, Corn Cob and Rice Straw for Ethanol Fermentation by Saccharomyces Cerevisiae. *Ann. Microbiol.* 2008, *58*, 219–225.

(174) Drapcho, C.M.; Nhuan, N.P. and Walker, T.H. Biofuels Engineering Process Technology. Mc Graw Hill Companies, Inc., 2008.

(175) Sivers, M.V. and Zacchi, G. A Techno-Economical Comparison of Three Processes for the Production of Ethanol from Pine. *Bioresource Technol.* 1995, *51*, 43–52.

(176) Root, D.F.; Saeman, J.F. and Harris, J.F. Kinetics of the Acid Catalyzed Conversion of Xylose to Furfural. *Forest Prod. J.* 1959, *158*, 165.

(177) Zeitsch, K.J. The Chemistry and Technology of Furfural and Its Many By-Products; Sugar Series. Elsevier: New York, 2000; *Vol. 13*.

(178) Esteghlalian, A.; Hashimoto, A.G.; Fenske, J.J. and Penner, M.H. Modeling and Optimization of the Dilute-Sulfuric-Acid Pretreatment of Corn Stover, Poplar and Switchgrass. *Bioresource Technol.* 1997, *59*, 129–136.

(179) Hinman, N.D.; Schell, D.J.; Riley, C.J.; Bergeron, P.W. and Walter, P.J. Preliminary Estimate of the Cost of Ethanol Production for SSF Technology. *Appl. Biochem. Biotechnol.* 1992, *34*, 639–649.

(180) Brennan, A.H.; Hoagland, W. and Schell, D.J. High Temperature Acid Hydrolysis of Biomass Using an Engineering-Scale Plug Flow Reactor: Result of Low Solids Testing. *Biotechnol. Bioeng. Symp.* 1986, *17*, 53–70.

(181) Converse, A.O.; Kwartneg, I.K.; Grethlein, H.E. and Ooshima, H. Kinetics of Thermochemical Pretreatment of Lignocellulosic Materials. *Appl. Biochem. Biotechnol.* 1989, *20*, 63–78.

(182) Brink, D.L. Method of Treating Biomass Material. U.S. Patent 5221357, 1993.

(183) Israilides, C.J.; Grant, G.A. and Han, Y.W. Sugar Level, Fermentability, and Acceptability of Straw Treated with Different Acids. *Appl. Environ. Microbiol.* 1978, *36*, 43–46.

(184) Goldstein, I.S.; Pereira, H.; Pittman, J.L.; Strouse, B.A. and Scaringelli, F.P. The Hydrolysis of Cellulose with Super Concentrated Hydrochloric Acid. *Biotechnol. Bioeng.* 1983, *13*, 17–25.

(185) Thompson, D.N.; Chen, H.C. and Grethlein, H.E. Comparison of Pretreatment Methods on the Basis of Available Surface Area. *Bioresource Technol.* 1991, *39*, 155–163.

(186) Ishizawa, C.I.; Davis, M.F.; Schell, D.F. and Hohnson, D.K. Porosity and Its Effect on the Digestibility of Dilute Sulfuric Acid Pretreated Corn Stover. *J. Agric. Food Chem.* 2007, *55*, 2575–2581.

(187) Lu, X.B.; Zhang, Y.M.; Yang, J. and Liang, Y. Enzymatic Hydrolysis of Corn Stover After Pretreatment with Dilute Sulfuric Acid. *Chem. Eng. Technol.* 2007, *30*, 938–944.

(188) Cara, C.; Ruiz, C.; Oliva, J.M.; Saez, F. and Castro, E. Conversion of Olive Tree Biomass into Fermentable Sugars by Dilute Acid Pretreatment and Enzymatic Saccharification. *Bioresource Technol.* 2008, *99*, 1869–1876.

(189) Selig, M.J.; Viamajala, S.; Decker, S.R.; Tuker, M.P. and Himmel, M.E. Deposition of Lignin Droplets Produced During Dilute Acid Pretreatment of Maize Stems Retards Enzymatic Hydrolysis of Cellulose. *Biotechnol. Prog.* 2007, *23*, 1333–1339.

(190) Wyman, C.E.; Dale, B.E.; Elander, R.T.; Holtzapple, M.; Ladisch, M.R. and Lee, Y.Y. Coordinated Development of Leading Biomass Pretreatment Technologies. *Bioresource Technol.* 2005, *96*, 1959–1966.

(191) Kumar, G.; Sen, B. and Lin, C.-Y. Pretreatment and Hydrolysis Methods for Recovery of Fermentable Sugars from De-oiled Jatropha Waste. *Bioresour. Technol.* 2013, *145*, 275–279.

(192) Adaganti, S.Y.; Yaliwal, V.S.; Kulkarni, B.M.; Desai, G.P. and Banapurmath, N.R. Factors Affecting Bioethanol Production from Lignocellulosic Biomass (Calliandra calothyrsus). *Waste Biomass Valori.* 2014, *5*, 963–971.

(193) Swaminathan, J.; Jayapal, P.R.; Asokan, M. and Ramasamy, D. Dilute Acid Pretreatment of Sweet Sorghum Stalk and Its Characterization. *Int. J. Chem Tech Res.* 2015, *8*, 589–598.

(194) Muktham, R.; Ball, A.S.; Bhargava, S.K. and Bankupalli, S. Bioethanol Production from Non-Edible De-oiled Pongamia Pinnata Seed Residue-Optimization of Acid Hydrolysis Followed by Fermentation. *Ind. Crop. Prod.* 2016, *94*, 490–497.

(195) Mohapatra, S.; Dandapat, S.J. and Thatoi, H. Physicochemical Characterization, Modelling and Optimization of Ultrasono Assisted Acid Pretreatment of Two Pennisetum sp. Using Taguchi and Artificial Neural Networking for Enhanced Delignification. *J. Environ. Manage.* 2017, *187*, 537–549.

(196) Gaspar, M.; Kalman, G. and Reczey, K. Corn Fibre as a Raw Material for Hemicellulose and Ethanol Production. *Process Biochem.* 2007, *42*, 1135–1139.

(197) Elshafei, A.M.; Vega, J.L.; Klasson, K.T.; Clausen, E.C. and Gaddy, J.L. The Saccharification of Corn Stover by Cellulase from Penicillin Funiculosum. *Bioresource Technol.* 1991, *35*, 73–80.

(198) Soto, M.L.; Dominguez, H.; Nunez, M.J. and Lema, J.M. Enzymatic Saccharification of Alkali-Treated Sunflower Hulls. *Bioresource Technol.* 1994, *49*, 53–59.

(199) Fox, D.J.; Gray, P.P.; Dunn, N.W. and Warwick, L.M. Comparison of Alkali and Steam (Acid) Pretreatments of Lignocellulosic Materials to Increase Enzymic Susceptibility: Evaluation Under Optimized Pretreatment Conditions. *J. Chem. Tech. Biotech.* 1989, *44*, 135–146.

(200) MacDonald, D.G.; Bakhshi, N.N.; Mathews, J.F.; Roychowdhury, A.; Bajpai, P. and Moo-Young, M. Alkali Treatment of Corn Stover to Improve Sugar Production by Enzymatic Hydrolysis. *Biotechnol. Bioeng.* 1983, *25*, 2067–2076

(201) Kim, S. and Holtzapple, M.T. Effect of Structural Features on Enzyme Digestibility of Corn Stover. *Bioresource Technol.* 2006, *97*, 583–591.

(202) Chang, V.S. and Holtzapple, M.T. Fundamental Factors Affecting Biomass Enzymatic Reactivity. *Appl. Biochem. Biotechnol.* 2000, *84*, 5–37.

(203) Lee, Y.H. and Fan, L.T. Kinetic Studies of Enzymatic Hydrolysis of Insoluble Cellulose: Analysis of the Initial Rates. *Biotechnol. Bioeng.* 1982, *24*, 2383–2406.

(204) Kong, F.; Engler, C.R. and Soltes, E.J. Effects of Cell Wall Acetate, Xylan Backbone, and Lignin on Enzymatic Hydrolysis of Aspen Wood. *Appl. Biochem. Biotechnol.* 1992, *34*, 23–35.

(205) Chang, V.S.; Nagwani, M. and Holtzapple, M.T. Lime Pretreatment of Crop Residues Bagasse and Wheat Straw. *Appl. Biochem. Biotechnol.* 1998, *74*, 135–159.

(206) Chang, V.S.; Nagwani, M.; Kim, C.H. and Holtzapple, M.T. Oxidative Lime Pretreatment of High-Lignin Biomass. *Appl. Biochem. Biotechnol.* 2001, *94*, 1–28.

(207) Chang, V.S.; Burr, B. and Holtzapple, M.T. Lime Pretreatment of Switch Grass. *Appl. Biochem. Biotechnol.* 1997, *63*, 3–19.

(208) Karr, W.E. and Holtzapple, M.T. Benefits from Tween during enzymic hydrolysis of corn stover. *Biotechnol. Bioeng.* 1998, *59*, 419–427.

(209) Karr, W.E. and Holtzapple, T. Using Lime Pretreatment to Facilitate the Enzymatic Hydrolysis of Corn Stover. *Biomass Bioenerg.* 2000, *18*, 189–199.

(210) Bjerre, A.B.; Olesen, A.B. and Fernqvist, T. Pretreatment of Wheat Straw Using Combined Wet Oxidation and Alkaline Hydrolysis Resulting in Convertible Cellulose and Hemicellulose. *Biotechnol. Bioeng.* 1996, *49*, 568–577.

(211) Chosdu, R.; Hilmy, N.E.; Erlinda, T.B. and Abbas, B. Radiation and Chemical Pretreatment of Cellulosic Waste. *Radiat. Phys. Chem.* 1993, *42*, 695–698.

(212) Hu, Z.; Wang, Y. and Wen, Z. Alkali (NaOH) Pretreatment of Switch Grass by Radio Frequency-Based Dielectric Heating. *Appl. Biochem. Biotechnol.* 2008, *148*, 71–81.

(213) Hu, Z. and Wen, Z. Enhancing Enzymatic Digestibility of Switch Grass by Microwave-Assisted Alkali Pretreatment. *Biochem. Eng. J.* 2008, *38*, 369–378.

(214) Iyer, V.; Wu, Z.W.; Kim, S.B. and Lee, Y.Y. Ammonia Recycled Percolation Process for Pretreatment of Herbaceous Biomass. *Appl. Biochem. Biotechnol.* 1996, *57*, 121–132.

(215) Kataria, R.; Ruhal, R.; Babu, R. and Ghosh, S. Saccharification of Alkali Treated Biomass of Kans Grass Contributes Higher Sugar in Contrast to Acid Treated Biomass. *Chem. Eng. J.* 2013, *230*, 36–47.

(216) Kataria, R. and Ghosh, S. Saccharification of Kans Grass Using Enzyme Mixture from Trichoderma Reesei for Bioethanol Production. *Bioresource Technol.* 2011, *102*, 9970–9975.

(217) Gao, K.; Boiano, S.; Marzocchella, A. and Rehmann, L. Cellulosic Butanol Production from Alkali-Pretreated Switchgrass (Panicum Virgatum) and Phragmites (Phragmites australis). *Bioresource Technol.* 2014, *174*, 176–181.

(218) Stoklosa, R.J. and Hodge, D.B. Fractionation and Improved Enzymatic Deconstruction of Hardwoods with Alkaline Delignification. *Bioenerg. Res.* 2015, *8*, 1224–1234.

(219) Siddhu, M.A.H.; Li, J.; Zhang, R.; Liu, J.; Ji, J.; He, Y.; Chen, C. and Liu, G. Potential of Black Liquor of Potassium Hydroxide to Pretreat Corn Stover for Biomethane Production. *Bioresources.* 2016, *11*, 4550–4563.

(220) Ling, Z.; Chen, S.; Zhang, X. and Xu, F. Exploring crystalline-Structural Variations of Cellulose During Alkaline Pretreatment for Enhanced Enzymatic Hydrolysis. *Bioresource Technol.* 2017, *224*, 611–617.

(221) Hon, D.N.S. and Shiraishi, N. Wood and Cellulosic Chemistry. Second Ed. Dekker; New York, 2001.

(222) Martel, P. and Gould, J.M. Cellulose Stability and Delignification after Alkaline Hydrogen-Peroxide Treatment of Straw. *J. Appl. Polym. Sci.* 1990, *39*, 707–714.

(223) Wei, C.J. and Cheng, C.Y. Effect of Hydrogen Peroxide Pretreatment on the Structural Features and the Enzymatic Hydrolysis of Rice Straw. *Biotechnol. Bioeng.* 1985, *27*, 1418–1426.

(224) Taniguchi, M.; Tanaka, M.; Matsuno, R. and Kamikubo, T. Evaluation of Chemical Pretreatment for Enzymatic Solubilization of Rice Straw. *Eur. J. Appl. Microbiol. Biotechnol.* 1982, *14*, 35–39.

(225) Toyama, N. and Ogawa, K. Sugar Production from Agricultural Woody Wastes by Saccharification with Trichoderma Viride Cellulase. *Biotechnol. Bioeng. Symp.* 1975, *5*, 225–244.

(226) Azzam, M. Pretreatment of Cane Bagasse with Alkaline Hydrogen Peroxide for Enzymatic Hydrolysis of Cellulose and Ethanol Fermentation. *J. Environ. Sci. Health B.* 1989, *24*, 421–433.

(227) Banerjee, G.; Car, S.; Liu, T.; Williams, D.L.; Meza, S.L.; Walton, J.D. and Hodge, D.B. Scale-up and Integration of Alkaline Hydrogen Peroxide Pretreatment, Enzymatic Hydrolysis, and Ethanolic Fermentation. *Biotechnol. Bioeng.* 2012, *109*, 922–931.

(228) Loow, Y.-L.; Wu, T.Y.; Lim, Y.S.; Tan, K.A.; Siow, L.F.; Md. Jahim, J. and Mohammad, A.W. Improvement of Xylose Recovery from the Stalks of Oil Palm Fronds Using Inorganic Salt and Oxidative Agent. *Energ. Convers. Manage.* 2017, *138*, 248–260.

(229) Botello, J.I.; Gilarranz, M.A.; Rodriguez, F. and Oliet, M. Preliminary Study on Products Distribution in Alcohol Pulping of Eucalyptus Globulus. *J. Chem. Technol. Biotechnol.* 1999, *74*, 141–148.

(230) Chum, H.L.; Johnson, D.K. and Black, S. Organosolv Pretreatment for Enzymatic Hydrolysis of Poplars: 1. Enzyme Hydrolysis of Cellulosic Residues. *Biotechnol. Bioeng.* 1988, *31*, 643–649.

(231) Thring, R.W.; Chorent, E. and Overend, R. Recovery of a Solvolytic Lignin: Effects of Spent Liquor/Acid Volume Ratio, Acid Concentration and Temperature. *Biomass.* 1990, *23*, 289–305.

(232) Sarkanen, K.V. Acid-Catalyzed Delignification of Lignocellulosics in Organic Solvents. *Prog. Biomass Convers.* 1980, *2*, 127–144.

(233) Pan, X.; Arato, C.; Gilkes, N.; Gregg, D.; Mabee, W.; Pye, K.; Xiao, Z.; Zhang, X. and Saddler, J. Biorefining of Softwoods Using Ethanol Organosolv Pulping: Preliminary Evaluation of Process Streams for Manufacture of Fuel-Grade Ethanol and Co-Products. *Biotechnol. Bioeng.* 2005, *90*, 473–481.

(234) Arato, C.; Pye, E.K. and Gjennestad, G. The Lignol Approach to Biorefining of Woody Biomass to Produce Ethanol and Chemicals. *Appl. Biochem. Biotechnol.* 2005, *121*, 871–882.
(235) Pan, X.; Gilkes, N.; Kadla, J.; Pye, K.; Saka, H.; Gregg, D.; Ehara, K.; Xie, D.; Lam, D. and Saddler, J. Bioconversion of Hybrid Poplar to Ethanol and Co-Products Using an Organosolv Fractionation Process: Optimization of Process Yields. *Biotechnol. Bioeng.* 2006, *94*, 851–861.
(236) Wildschut, J.; Smit, A.T.; Reith, J.H. and Huijgen, W.J.J. Ethanol-Based Organosolv Fractionation of Wheat Straw for the Production of Lignin and Enzymatically Digestible Cellulose. *Bioresource Technol.* 2013, *135*, 2013, 58–66.
(237) Bouxin, F.P.; David Jackson, S. and Jarvis, M.C. Organosolv Pretreatment of Sitka Spruce Wood: Conversion of Hemicelluloses to Ethyl Glycosides. *Bioresource Technol.* 2014, *151*, 441–444.
(238) Gandolfi, S.; Ottolina, G.; Consonni, R.; Riva, S. and Patel, I. Fractionation of Hemp Hurds by Organosolv Pretreatment and Its Effect on Production of Lignin and Sugars. *Chem.Sus.Chem.* 2014, *7*, 1991–1999.
(239) Kabir, M.M.; Rajendran, K.; Taherzadeh, M.J. and Sárvári Horváth, I. Experimental and Economical Evaluation of Bioconversion of Forest Residues to Biogas Using Organosolv Pretreatment. *Bioresource Technol.* 2015, *178*, 201–208.
(240) Chen, H.; Zhao, J.; Hu, T.; Zhao, X. and Liu, D. A Comparison of Several Organosolv Pretreatments for Improving the Enzymatic Hydrolysis of Wheat Straw: Substrate Digestibility, Fermentability and Structural Features. *Appl. Energ.* 2015, *150*, 224–232.
(241) Nitsos, C.; Stoklosa, R.; Karnaouri, A.; Voros, D.; Lange, H.; Hodge, D.; Crestini, C.; Rova, U. and Christakopoulos, P. Isolation and Characterization of Organosolv and Alkaline Lignins from Hardwood and Softwood Biomass. *ACS Sustain. Chem. Eng.* 2016, *4*, 5181–5193.
(242) Asadi, N. and Zilouei, H. Optimization of Organosolv Pretreatment of Rice Straw for Enhanced Biohydrogen Production Using Enterobacter Aerogenes. *Bioresource Technol.* 2017, *227*, 335–344.
(243) Chundawat, S.P.S.; Bellesia, G.; Uppugundla, N.; da Costa Sousa, L.; Gao, D.; Cheh, A.M.; Agarwal, U.P.; Bianchetti, C.M.; Phillips, G.N.; Langan, P.; Balan, V.; Gnanakaran, S. and Dale, B.E. Restructuring the Crystalline Cellulose Hydrogen Bond Network Enhances Its Depolymerization Rate. *J. Am. Chem. Soc.* 2011, *133*, 11163–11174.
(244) Gao, D.; Chundawat, S.P.S.; Sethi, A.; Balan, V.; Gnanakaran, S. and Dale, B.E. Increased Enzyme Binding to Substrate Is Not

Necessary for More Efficient Cellulose Hydrolysis. *Proc. Natl. Acad. Sci. U.S.A.* 2013, *110*, 10922–10927.

(245) Belokurova, O.A.; Kirillova, M.N.; Shcheglova, T.L. and Mel'nikov, B.N. Use of Liquid Ammonia to Increase the Quality of Textiles Made of Hydrated Cellulose Fibres. *Fibre Chem.* 1996, *28*, 259–263.

(246) Rousselle, M.A.; Nelson, M.L.; Hassenboehler, C.B. and Legendre, D.C. Liquid-Ammonia and Caustic Mercerization of Cotton Fibers: Changes in Fine Structure and Mechanical Properties. *Text. Res. J.* 1976, *46*, 304–310.

(247) Lewin, M. and Roldan, L.G. The Effect of Liquid Anhydrous Ammonia on the Structure and Morphology of Cotton Cellulose. *J. Polym. Sci., Part C: Polym. Symp.* 1971, *36*, 213–229.

(248) Barry, A.J.; Peterson, F.C. and King, A.J. X-Ray Studies of Reactions of Cellulose in Non-Aqueous Systems. I. Interaction of Cellulose and Liquid Ammonia. *J. Am. Chem. Soc.* 1936, *58*, 333–337.

(249) Chundawat, S.P.S.; Donohoe, B.S.; Sousa, L.D.; Elder, T.; Agarwal, U.P.; Lu, F.C.; Ralph, J.; Himmel, M.E.; Balan, V. and Dale, B.E. Multi Scale Visualization and Characterization of Lignocellulosic Plant Cell Wall Deconstruction during Thermochemical Pretreatment. *Energy Environ. Sci.* 2011, *4*, 973–984.

(250) Costa Sousa, da. L.; Jin, M.; Chundawat, S.P.S.; Bokade, V.; Tang, X.; Azarpira, A.; Lu, F.; Avci, U.; Humpula, J.; Uppugundla, N.; Gunawan, C.; Pattathil, S.; Cheh, A.M.; Kothari, N.; Kumar, R.; Ralph, J.; Michael G. Hahn, M.G.; Wyman, C.E.; Singh, S.; Simmons, B.A.; Bruce E. Dale, B.A. and Balan, V. Next-Generation Ammonia Pretreatment Enhances Cellulosic Biofuel Production. *Energy Environ. Sci.* 2016, *9*, 1215–1223.

(251) Campbell, T.J.; Teymouri, F.; Bals, B.; Glassbrook, J.; Nielson, C.D.; Videto, J. and A. Packed Bed Ammonia Fiber Expansion Reactor System for Pretreatment of Agricultural Residues at Regional Depots. *Biofuels.* 2013, *4*, 23–34.

(252) Schuerch, C. Plasticizing Wood with Liquid Ammonia. *J. Ind. Eng. Chem.* 1963, *55*, 39.

(253) Chundawat, S.P.S.; Vismeh, R.; Sharma, L.N.; Humpula, J.F.; da Costa Sousa, L.; Chambliss, C.K.; Jones, A.D.; Balan, V. and Dale, B.E. Multifaceted Characterization of Cell Wall Decomposition Products Formed During Ammonia Fiber Expansion (AFEX) and Dilute Acid Based Pretreatments. *Bioresour. Technol.* 2010, *101*, 8429–8438.

(254) Pattathil, S.; Hahn, M.G.; Dale, B.E. and Chundawat, S.P.S. Insights Into Plant Cell Wall Structure, Architecture, and Integrity

Using Glycome Profiling of Native and AFEX[TM] Pretreated Biomass. *J. Exp. Bot.* 2015, *66*, 4279–4294.

(255) Pan, X.J. Role of Functional Groups in Lignin Inhibition of Enzymatic Hydrolysis of Cellulose to Glucose. *J. Biobased Mater. Bio.* 2008, *2*, 25–32.

(256) E. Palmqvist, E. and Hahn-Hagerdal, B. Fermentation of Lignocellulosic Hydrolysates. II: Inhibitors and Mechanisms of Inhibition. *Bioresour. Technol.* 2000, *74*, 25–33.

(257) Hayes, D.J. An Examination of Biorefining Processes, Catalysts and Challenges. *Catal. Today.* 2009, *145*, 138–151.

(258) Li, Q.; He, Y.C.; Xian, M.; Jun, G.; Xu, X.; Yang, J.M. and Li, L.Z. Improving Enzymatic Hydrolysis of Wheat Straw Using Ionic Liquid 1-Ethyl-3-Methyl Imidazolium Diethyl Phosphate Pretreatment. *Bioresour. Technol.* 2009, *100*, 3570–3575.

(259) Lee, S.H.; Doherty, T.V.; Linhardt, R.J. and Dordick, J.S. Ionic Liquid-Mediated Selective Extraction of Lignin from Wood Leading to Enhanced Enzymatic Cellulose Hydrolysis. *Biotechnol. Bioeng.* 2009, *102*, 1368–1376.

(260) Yang, B. and Wyman, C.E. Pretreatment: The Key to Unlocking Low-Cost Cellulosic Ethanol. *Biofuels Bioprod. Bior.* 2008, *2*, 26–40.

(261) Zhao, H.; Jones, C.L.; Baker, G.A.; Xia, S.; Olubajo, O. and Person, V.N. Regenerating Cellulose from Ionic Liquids for an Accelerated Enzymatic Hydrolysis. *J. Biotechnol.* 2009, *139*, 47–54.

(262) Shi, J.; Gladden, J.M.; Sathitsuksanoh, N.; Kambam, P.; Sandoval, L.; Mitra, D.; Zhang, S.; George, A.; Singer, S.W.; Simmons, B.A. and Singh, S. One-Pot Ionic Liquid Pretreatment and Saccharification of Switchgrass. *Green Chem.* 2013, *15*, 2579–2589.

(263) Verdía, P.; Brandt, A.; Hallett, J.P.; Ray, M.J. and Welton, T. Fractionation of Lignocellulosic Biomass with the Ionic Liquid 1-Butylimidazolium Hydrogen Sulfate. *Green Chem.* 2014, *16*, 1617–1627.

(264) George, A.; Brandt, A.; Tran, K.; Zahari, S.M.S.N.S.; Klein-Marcuschamer, D.; Sun, N.; Sathitsuksanoh, N.; Shi, J.; Stavila, V.; Parthasarathi, R.; Singh, S.; Holmes, B.M.; Welton, T.; Simmons, B.A. and Hallett, J.P. Design of Low-Cost Ionic Liquids for Lignocellulosic Biomass Pretreatment. *Green Chem.* 2015, *17*, 1728–1734.

(265) Allison, B.J.; CAidiz, J.C.; Karuna, N.; Jeoh, T. and Simmons, C.W. The Effect of Ionic Liquid Pretreatment on the Bioconversion of Tomato Processing Waste to Fermentable Sugars and Biogas. *Appl. Biochem. Biotechnol.* 2016, *179*, 1227–1247.

(266) Uju, G.M. and Kamiya, N. Powerful Peracetic Acid-Ionic Liquid Pretreatment Process for the Efficient Chemical Hydrolysis of Lignocellulosic Biomass. *Bioresource Technol.* 2016, *214*, 487–495.

(267) Chang, K.-L.; Chen, X.-M.; Wang, X.-Q.; Han, Y.-J.; Potprommanee, L.; Liu, J.-Y.; Liao, Y.-L.; Ning, X.-A.; Sun, S.-Y. and Huang, Q. Impact of Surfactant Type for Ionic Liquid Pretreatment on Enhancing Delignification of Rice Straw. *Bioresource Technol.* 2017, *227*, 388–392.

(268) Kassaye, S.; Pant, K.K. and Jain, S. Hydrolysis of Cellulosic Bamboo Biomass into Reducing Sugars via a Combined Alkaline Solution and Ionic Liquid Pretreament Steps. *Renew. Energ.* 2017, *104*, 177–184.

(269) Parthasarathi, R.; Sun, J.; Dutta, T.; Sun, N.; Pattathil, S.; Murthy, K.N.V.; Peralta, A.G., Simmons, B.A. and Singh, S. Activation of Lignocellulosic Biomass for Higher Sugar Yields Using Aqueous Ionic Liquid at Low Severity Process Conditions. *Biotechn. Biofuels.* 2016, *9*, 160.

(270) Pepper, J.M. and Lee, Y.W. Lignin and Related Compounds. I. A Comparative Study of Catalysts for Lignin Hydrogenolysis. *Can. J. Chem.* 1969, *47*, 723.

(271) Sturgeon, M.R.; Kim, S.; Lawrence, K.; Paton, R.S.; Chmely, S.C.; Nimlos, M.; Foust, T.D. and Beckham, G.T. A Mechanistic Investigation of Acid-Catalyzed Cleavage of Aryl-Ether Linkages: Implications for Lignin Depolymerization in Acidic Environments. *ACS Sustain. Chem. Eng.* 2014, *2*, 472–485.

(272) Deuss, P.J.; Scott, M.; Tran, F.; Westwood, N.J.; de Vries, J.G. and Barta, K. Aromatic Monomers by in Situ Conversion of Reactive Intermediates in the Acid-Catalyzed Depolymerization of Lignin. *J. Am. Chem. Soc.* 2015, *137*, 7456–7467.

(273) Constant, S.; Wienk, H.L.J.; Frissen, A.E.; Peinder, P.D.; Boelens, R.; van Es, D.S.; Grisel, R.J.H.; Weckhuysen, B.M.; Huijgen, W.J.J.; Gosselink, R.J.A. and Bruijnincx, P.C.A. New Insights into The Structure and Composition of Technical Lignins: A Comparative Characterization Study. *Green Chem.* 2016, *18*, 2651–2665.

(274) Ragauskas, A.J.; Beckham, G.T.; Biddy, M.J.; Chandra, R.; Chen, F.; Davis, M.F.; Davison, B.H.; Dixon, R.A.; Gilna, P.; Keller, M.; Langan, P.; Naskar, A.K.; Saddler, J.N.; Tschaplinski, T.J.; Tuskan, G.A.; Wyman, C.E. Lignin Valorization: Improving Lignin Processing in the Biorefinery. *Science.* 2014, *344*, 6185.

(275) Zakzeski, J.; Bruijnincx, P.C.A.; Jongerius, A.L. and Weckhuysen, B.M. The Catalytic Valorization of Lignin for the Production of Renewable Chemicals. *Chem. Rev.* 2010, *110*, 3552–3599.

(276) Beckham, G.T.; Johnson, C.W.; Karp, E.M.; Salvachúa, D. and Vardon, D.R. Opportunities and Challenges in Biological Lignin Valorization. *Curr. Opin. Biotechnol.* 2016, *42*, 40–53.

(277) Rinaldi, R.; Jastrzebski, R.; Clough, M.T.; Ralph, J.; Kennema, M.; Bruijnincx, P.C. and Weckhuysen, B.M. Paving the Way for Lignin Valorisation: Recent Advances in Bioengineering, Biorefining and Catalysis. *Angew. Chem. Int. Ed.*2016, *55*, 8164–8215.

(278) Schutyser, W.; Van den Bosch, S.; Renders, T.; De Boe, T.; Koelewijn, S.F.; Dewaele, A.; Ennaert, T.; Verkinderen, O.; Goderis, B.; Courtin, C.M. and Sels, B.F. Influence of Biobased Solvents on the Catalytic Reductive Fractionation of Birch Wood. *Green Chem.* 2015, *17*, 5035–5045.

(279) Van den Bosch, S.; Schutyser, W.; Vanholme, R.; Driessen, T.; Koelewijn, S.F.; Renders, T.; De Meester, B.; Huijgen, W.J.J.; Dehaen, W.; Courtin, C.M.; Lagrain, B.; Boerjan, W. and Sels, B.F. Reductive Lignocellulose Fractionation into Soluble Lignin Derived Phenolic Monomers and Dimers and Processable Carbohydrate Pulps. *Energy Environ. Sci.* 2015, *8*, 1748–1763.

(280) Parsell, T.H.; Owen, B.C.; Klein, I.; Jarrell, T.M.; Marcum, C.L.; Haupert, L.J.; Amundson, L.M.; Kenttamaa, H.I.; Ribeiro, F.; Miller, J.T. and Abu-Omar, M.M. Cleavage and Hydrodeoxygenation (HDO) of C-O bonds Relevant to Lignin Conversion Using Pd/Zn Synergistic Catalysis. *Chem. Sci.* 2013, *4*, 806–813.

(281) Galkin, M.V.; Sawadjoon, S.; Rohde, V.; Dawange, M. and Samec, J.S.M. Mild Heterogeneous Palladium-Catalyzed Cleavage of β-O-4-Ether Linkages of Lignin Model Compounds and Native Lignin in Air. *Chem. Cat. Chem.* 2014, *6*, 179–184.

(282) Song, Q.; Wang, F.; Cai, J.; Wang, Y.; Zhang, J.; Yu, W. and Xu, J. Lignin Depolymerization (LDP) in Alcohol over Nickel Based Catalysts via a Fragmentation-Hydrogenolysis Process. *Energy Environ. Sci.* 2013, *6*, 994–1007.

(283) Ferrini, P. and Rinaldi, R. Catalytic Biorefining of Plant Biomass to Non-Pyrolytic Lignin Bio-Oil and Carbohydrates through Hydrogen Transfer Reactions. *Angew. Chem. Int. Ed.* 2014, *53*, 8634–8639.

(284) Yan, N.; Zhao, C.; Dyson, P.J.; Wang, C.; Liu, L.-t. and Kou, Y. Selective Degradation of Wood Lignin over Noble-Metal Catalysts in a Two-Step Process. *Chem. Sus. Chem.* 2008, *1*, 626–629.

(285) Grabber, J.H.; Quideau, S. and Ralph, J. p-Coumaroylated Syringyl Units in Maize Lignin: Implications for β-Ether Cleavage by Thioacidolysis. *Phytochemistry.* 1996, *43*, 1189–1194.

(286) Ralph, J. Hydroxycinnamates in Lignification. *Phytochem. Rev.* 2010, *9*, 65–83.

(287) Ralph, J.; Bunzel, M.; Marita, J.M.; Hatfield, R.D.; Lu, F.; Kim, H.; Schatz, P.F.; Grabber, J.H. and Steinhart, H. Peroxidase Dependent Cross Linking Reactions of Phydroxycinnamates in Plant Cell Walls. *Phytochem. Rev.* 2004, *3*, 79–96.

(288) Ralph, J.; Hatfield, R.D.; Quideau, S.; Helm, R.F.; Grabber, J.H. and Jung, H.J.G. Pathway of p-Coumaric Acid Incorporation into Maize Lignin As Revealed by NMR. *J. Am. Chem. Soc.* 1994, *116*, 9448–9456.

(289) Lu, F. and Ralph, J. Detection and Determination of p-Coumaroylated Units in Lignins. *J. Agric. Food Chem.* 1999, *47*, 1988–1992.

(290) Luo, H.; Klein, I.M.; Jiang, Y.; Zhu, H.; Liu, B.; Kenttämaa, H.I. and Abu-Omar, M.M. Total Utilization of Miscanthus Biomass, Lignin and Carbohydrates, Using Earth Abundant Nickel Catalyst. ACS Sustain. *Chem. Eng.* 2016, *4*, 2316–2322.

(291) Langholtz, M.H. and Eaton, B.J.S.L.M. Billion Ton Report: Advancing Domestic Resources for a Thriving Bioeconomy, Volume 1: Economic Availability of Feed Stocks. U.S. Department of Energy, *Vol. 1*, 2016.

(292) Anderson, E.M.; Katahira, R.; Reed, M.; Resch, M.G.; Karp, E.M.; Beckham, G.T. and Roman-Leshkov, Y. Reductive Catalytic Fractionation of Corn Stover Lignin. *ACS Sustain. Chem. Eng.* 2016, *4*, 6940–6950.

(293) Perlack, R.D.; Stokes, S.B.J. U.S. Billion-Ton update: Biomass Supply for a Bioenergy and Bioproducts Industry. U.S. Department of Energy, 2011.

(294) Renders, T.; Schutyser, W.; Van den Bosch, S.; Koelewijn, S.-F.; Vangeel, T.; Courtin, C.M. and Sels, B.F. Influence of Acidic (H_3PO_4) and Alkaline (NaOH) Additives on the Catalytic Reductive Fractionation of Lignocellulose. *ACS Catal.* 2016, *6*, 2055–2066.

(295) Abbott, A.P.; Capper, G.; Davies, D.L.; Rasheed, R.K. and Tambyrajah, V. Novel Solvent Properties of Choline Chloride/ Urea Mixtures. *Chem. Commun.* 2003, *1*, 70–71.

(296) Smith, E.L.; Abbott, A.P. and Ryder, K.S. Deep Eutectic Solvents (DESs) and Their Applications. *Chem. Rev.* 2014, *114*, 11060–11082.

(297) Carriazo, D.; Serrano, M.C.; Gutierrez, M.C.; Ferrer, M.L. and del Monte, F. Deep-Eutectic Solvents Playing Multiple Roles in the Synthesis of Polymers and Related Materials. *Chem. Soc. Rev.* 2012, *41*, 4996–5014.

(298) Zhang, Q.; Vigier, K.D.O.; Royer, S. and Jerome, F. Deep Eutectic Solvents: Syntheses, Properties and Applications. *Chem. Soc. Rev.* 2012, *41*, 7108–7146.

(299) Wang, J.; Wang, P.; Wang, Q.; Mou, H.; Cao, B.; Yu, D.; Wang, D.; Zhang, S. and Mu, T. Low Temperature Electrochemical Deposition of Aluminum in Organic Bases/Thiourea-Based Deep Eutectic Solvents. *ACS Sustain. Chem. Eng.* 2018, *6*, 15480–15486.

(300) Zdanowicz, M.; Wilpiszewska, K. and Spychaj, T. Deep Eutectic Solvents for Polysaccharides Processing. A Review. *Carbohydr. Polym.* 2018, *200*, 361–380.

(301) van Oschm D.J.G.P.; Kollau, L.J.B.M.; van den Bruinhorst, A.; Asikainen, S.; Rocha, M.A.A. and Kroon, M.C. Ionic Liquids and Deep Eutectic Solvents for Lignocellulosic Biomass Fractionation. *Phys. Chem. Chem. Phys.* 2017, *19*, 2636–2665.

(302) Tang, X.; Zuo, M.; Li, Z.; Liu, H.; Xiong, C.; Zeng, X.; Sun, Y.; Hu, L.; Liu, S.; Lei, T. and Lin, L. Green Processing of Lignocellulosic Biomass and Its Derivatives in Deep Eutectic Solvents. *Chem SusC hem.* 2017, *10*, 2696–2706.

(303) Loow, Y.-L.; New, E.K.; Yang, G.H.; Ang, L.Y.; Foo, L.Y.W. and Wu, T.Y. Potential Use of Deep Eutectic Solvents to Facilitate Lignocellulosic Biomass Utilization and Conversion. *Cellulose.* 2017, *24*, 3591–3618.

(304) Vigier, K.D.O.; Chatel, G. and Jerome, F. Contribution of Deep Eutectic Solvents for Biomass Processing: Opportunities, Challenges, and Limitations. *Chemcatchem.* 2015, *7*, 1250–1260.

(305) Turbak, A.F.; Hammer, R.B.; Davies, R.E. and Hergert, H.L. Recent Developments in Organic Synthesis by Electrolysis. *Chemtech.* 1980, *10*, 51–57.

(306) Swatloski, R.P.; Spear, S.K.; Holbrey, J.D. and Rogers, R.D. Dissolution of Cellose With Ionic Liquids. *J. Am. Chem. Soc.* 2002, *124* 4974–4975.

(307) Ren, H.; Chen, C.; Wang, Q.; Zhao, D. and Guo, S. *Bioresources.* 2016, *11*, 5435–5451.

(308) Laitinen, O.; Ojala, J.; Sirvio, J.A. and Liimatainen, H. The Properties of Choline Chloride-based Deep Eutectic Solvents and Their Performance in the Dissolution of Cellulose. *Cellulose.* 2017, *24*, 1679–1689.

(309) Liu, Y.; Guo, B.; Xia, Q.; Meng, J.; Chen, W.; Liu, S.; Wang, Q.; Liu, Y.; Li, J. and Yu, H. Efficient Cleavage of Strong Hydrogen Bonds in Cotton by Deep Eutectic Solvents and Facile Fabrication of Cellulose Nanocrystals in High Yields. *ACS Sustain. Chem. Eng.* 2017, *5*, 7623–7631.

(310) Taniguchi, M.; Suzuki, H.; Watanabe, D.; Sakai, K.; Hoshino, K. and Tanaka, T. Evaluation of Pretreatment with Pleurotus

Ostreatus for Enzymatic Hydrolysis of Rice Straw. *J. Biosci. Bioeng.* 2005, *100*, 637–643.

(311) Okano, K.; Kitagaw, M.; Sasaki, Y. and Watanabe, T. Conversion of Japanese Red Cedar (*Cryptomeria japonica*) into a Feed for Ruminants by White-Rot Basidiomycetes. *Animal Feed Sci. Technol.* 2005, *120*, 235–243.

(312) Itoh, H.; Wada, M.; Honda, Y.; Kuwahara, M. and Watanabe, T. Bio-Organosolve Pretreatments for Simultaneous Saccharification and Fermentation of Beech Wood by Ethanolysis and White Rot Fungi. *J. Biotechnol.* 2003, *103*, 273–280.

(313) Balan, V.; Souca, L.D.C.; Chundawat, S.P.S.; Vismeh, R.; Jones, A.D.; Dale, B.E. Mushroom Spent Straw: A Potential Substrate for an Ethanol-Based Biorefinery. *J. Ind. Microbiol. Biotechnol.* 2008, *35*, 293–301.

(314) Zhao, L.; Cao, G.-L.; Wang, A.-J.; Ren, H.-Y.; Dong, D.; Liu, Z.-N.; Guan, X.-Y.; Xu, C.-J. and Ren, N.-Q. Fungal Pretreatment of Cornstalk with Phanerochaete Chrysosporium for Enhancing Enzymatic Saccharification and Hydrogen Production. *Bioresource Technol.* 2012, *114*, 365–369.

(315) Liu, J.; Wang, M.L.; Tonnis, B.; Habteselassie, M.; Liao, X. and Huang, Q. Fungal Pretreatment of Switchgrass for Improved Saccharification and Simultaneous Enzyme Production. *Bioresource Technol.* 2013, *135*, 39–45.

(316) Cianchetta, S.; Di Maggio, B.; Burzi, P.L. and Galletti, S. Evaluation of Selected White-Rot Fungal Isolates for Improving the Sugar Yield from Wheat Straw. *Appl. Biochem. Biotech.* 2014, *173*, 609–623.

(317) Wen, B.; Yuan, X.; Li, Q.X.; Liu, J.; Ren, J.; Wang, X. and Cui, Z. Comparison and Evaluation of Concurrent Saccharification and Anaerobic Digestion of Napier Grass after Pretreatment by Three Microbial Consortia. *Bioresource Technol.* 2015, *175*, 102–111.

(318) Kumari, S. and Das, D. Biologically Pretreated Sugarcane Top as a Potential Raw Material for the Enhancement of Gaseous Energy Recovery by Two Stage Biohythane Process. *Bioresource Technol.* 2016, *218*, 1090–1097.

(319) Kavitha, S.; Subbulakshmi, P.; Rajesh Banu, J.; Gobi, M. and Tae Yeom, I., Enhancement of Biogas Production from Microalgal Biomass through Cellulolytic Bacterial Pretreatment. *Bioresource Technol.* 2017, *233*, 34–43.

(320) Lynch, P.T.; Davey, M.R. Electrical Manipulation of Cells; Chapman and Hall: New York, 1996.

(321) Zimmermann, U.; Neil, G.A. Electromanipulation of Cells; CRC Press: Boca Raton, FL, 1996.

(322) Vangerbasch, A.; Heinz, V. and Knorr, D. Effects of Pulsed Electric Fields on Cell Membranes in Real Food Systems. *Innov. Food Sci. Emerg. Technol.* 2000, *1*, 135–149.

(323) Jayaram, S.; Catle, G.S.P. and Margaritis, A. The Effect of High Field DC Pulse and Liquid Medium Conductivity on Survivability of Lactobacillus Brevis. *Appl. Microbiol. Biotechnol.* 1993, *40*, 117–122.

(324) Giner, G.; Gimeno, V.; Barbosa-Canovas, G.V. and Martin, O. Effects of Pulsed Electric Field Processing on Apples and Pear Polyphenoloxidases. *Food Sci. Technol. Int.* 2001, *7*, 339–345.

(325) Taiwo, K.A.; Angerbasch, A.; Ade-Omowaye, B.I.O. and Knorr, D. Effects of Pretreatments on the Diffusion Kinetics and Some Quality Parameters of Osmotically Dehydrated Apple Slices. *J. Agric. Food Chem.* 2001, *49*, 2804–2811.

(326) Eshtiaghi, M.N. and Knorr, D. High Electric Field Pulse Treatment: Potential for Sugar Beet Processing. *J. Food Eng.* 2002, *52*, 265–272.

(327) Bazhal, M.I.; Lebvoka, N.I. and Vorobiev, E. Pulsed Electric Field Treatment of Apple Tissue during Compression for Juice Extraction. *J. Food Eng.* 2001, *50*, 129–139.

(328) Zhu, S.; Wu, Y.; Zhao, Y.; Tu, S. and Xue, Y. Fed Batch Simultaneous Saccharification and Fermentation of Microwave/Acid/Alkali/H_2O_2 Pretreated Rice Straw for Production of Ethanol. *Chem. Eng.Commun.* 2006, *193*, 639–648

(329) Lu, Z.X. and Minoru, K. Effect of Radiation Pretreatment on Enzymatic Hydrolysis of Rice Straw with Low Concentrations of Alkali Solution. *Bioresour. Technol.* 1993, *43*, 13–17.

(330) Jin, S. and Chen, H. Superfine Grinding of Steam-Exploded Rice Straw and Its Enzymatic Hydrolysis. *Biochem. Eng. J.* 2006, *30*, 225–230.

(331) Zhu, Z.; Macquarrie, D.J.; Simister, R.; Gomez, L.D. and McQueen-Mason, S.J. Microwave Assisted Chemical Pretreatment of Miscanthus under Different Temperature Regimes. *Sustain. Chem. Process.* 2015, *3*, 15.

(332) Verma, P.; Watanabe, T.; Honda, Y. and Watanabe, T. Microwave-Assisted Pretreatment of Woody Biomass with Ammonium Molybdate Activated by H_2O_2. *Bioresource Technol.* 2011, *102*, 3941–3945.

(333) Ethaib, S.; Omar, R.; Kamal, S.M.M.; Biak, D.R.A. and Harun, S.S.M.Y. Microwave-Assisted Pretreatment of Sago Palm Bark. *J. Wood Chem. Technol.* 2017, *37*, 26–42.

(334) Bala, G.S.; Chennuru, R.; Mahapatra, S. and Vinu, R. Effect of Alkaline Ultrasonic Pretreatment on Crystalline Morphology and

Enzymatic Hydrolysis of Cellulose. *Cellulose.* 2016, *23*, 1723–1740.

(335) Ogg, C. *Environmental Challenges Associated with Corn Ethanol Production.* Biofuels, Food, and Feed Tradeoffs Conference organized by the Farm Foundation: St. Louis, Missouri, April 12–13, 2007.

(336) Curran, L.; Trigg, S.N.; McDonald, A.K.; Astiani, D.; Hardiono, Y.M.;Siregar, P.; Caniago, I. and Kasischke, E. Lowland Forest Loss in Protected Areas of Indonesian Borneo. *Science.* 2004, *313*, 1000–1003.

(337) Brazilian Foundation for Sustainable Development (FBDS). Outcomes and Recommendations of Workshop on Agro-energy Expansion and Its Impact on Brazilian Natural Ecosystems. 2007.

(338) Rajagopal, D. Rethinking Current Strategies for Biofuel Production in India. International Conference on Linkages in Water and Energy in Developing Countries.Organized by IWMI and FAO, ICRISAT: Hyderabad, India, January 29–30, 2007.

(339) Gundimeda, H. How Sustainable Is the Sustainable Development Objective of CDM in Developing Countries like India? *Forest Policy Econ.* 2004, *6*, 329–343.

(340) Treguer, D.; J. Sourie, J. The Impact of Biofuel Production on Farm Jobs and Income: The French Case. The 96th EAAE seminar in Tanikon: Switzerland, January 2006.

(341) Elobeid, A.; Tokgoz, S. Removal of U. S. Ethanol Domestic and Trade Distortions: Impact on U. S. and Brazilian Ethanol Markets. Working Paper 06-WP 427, Center for Agricultural and Rural Development, Iowa State University, Ames, Iowa, 2006.

(342) Banse, M.; Tabeau, A.; Woltjer, G.; vanMeijl. H. Impact of EU Biofuel Policies on World Agricultural and Food Markets. The GTAP Conference, Purdue University: Indiana, 2007.

(343) Fischer, G. and Schrattenholzer. L. Global Bioenergy Potentials through 2050. *Biomass Bioenerg.* 2001, *20*, 151–159.

(344) Ugarte, D.G. de la Torre. Developing Bioenergy: Economic and Social Issues, in Bioenergy and Agriculture: Promises and Challenges. International Food Policy Research Institute 2020 Focus No. 14, 2006.

(345) Organization for Economic Cooperation and Development (OECD). Agricultural Market Impacts of Future Growth in the Production of Biofuels. OECD Papers 2006, *6*, 1–57.

(346) Berndes, G.; Hoogwijk, M. and van den Broek, R. The Contribution of Biomass in the Future Global Energy Supply: A Review of 17 Studies. *Biomass Bioenerg.* 2003, *25*, 1–28.

(347) Greene, N.; Celik, F.E.; Dale, B.; Jackson, M.; Jayawardhana, K.; Jin, H.; Larson. E.; Laser, M.; Lynd, K.; MacKenzie, D.; Jason, M.;

McBride, J.; McLaughlin, S.; Saccardi, D. *Growing Energy: How Biofuels Can Help End America's Oil Dependence.* National Resource Defense Council (NRDC) Report, 2004.

(348) Waggoner, P. and Ausubel, J. How Much Will Feeding More and Wealthier People Encroach on Forests? *Popul. Dev. Rev.* 2001, *27*, 239–257.

(349) Waggoner, P.E. How Much Land Can Ten Billion People Spare for Nature? *Daedalus.* Summer 1996, *125*, 3.

(350) Kammen, D.; Kapadia, K.; Fripp, M. Putting Renewables to Work: How Many Jobs Can the Clean Energy Industry Generate. Report of the Renewable and Appropriate Energy Laboratory, Energy and Resources Group/Goldman School of Public Policy at University of California, Berkeley, USA, UCL Publishers, April, 2004.

(351) Kammen, D.M. Bioenergy in Developing Countries: Experiences and Prospects, in Bioenergy and Agriculture: Promises and Challenges. International Food Policy Research Institute 2020 Focus No. 14, 2006.

(352) Hazell, P.; Pachauri, R.K. Bioenergy and Agriculture: Promises and Challenges. International Food Policy Research Institute 2020 Focus No. 14. International Energy Agency (IEA). IEA Global Renewable Fact Sheet 2006.

(353) Pandey, A.; Soccol, C.R.; Nigam, P.; Soccol, V.T.; Vandenberghe, L.P.S. and Mohan, R. Biotechnological Potential of Agro-Industrial Residues. II: Cassava Bagasse. *Bioresource Technol.* 2000, *74*, 81–87.

(354) Runge, C.; Senauer, B. How Biofuels Could Starve the Poor. Foreign Affairs 2007.

(355) Msangi, S.; Sulser, T.; Rosegrant, M.; Valmonte-Santos, R.; Ringler, C. Global Scenarios for Biofuels: Impacts and Implications. International Food Policy Research Institute (IFPRI), 2006.

(356) van Dam, J.; Junginger, M.; Faaij, A.; Jürgens, I.; Best, G.; Fritsche, U. Overview of Recent Developments in Sustainable Biomass Certification. Biomass Bioenerg. 2007, 32, 749–780.

(357) Fearnside, P. Soybean Cultivation as a Threat to the Environment in Brazil. *Environ. Conserv.* 2002, *28*, 23–38.

(358) Giampietro, M.; Ulgiati, S. and Pimentel, D. Feasibility of Large-Scale Biofuel Production. *Biosci.* 1997, *47*, 587–600.

(359) Smith, K.R. Biofuels, Air Pollution, and Health: A Global Review. Plenum: New York, 1987.

(360) Smith, K. and Mehta. S. The Burden of Disease from Indoor Air Pollution in Developing Countries: Comparison of Estimates. *Int. J. Hyg. Environ Health.* 2003, *206*, 279–289.

(361) Ravindranath, N.H.; Hall, D.O. Biomass, Energy, and Environment: A Developing Country Perspective from India. Oxford University Press: Oxford, 1995.
(362) Barnes, D.F. and Floor, W.M. Rural Energy in Developing Countries: A Challenge for Economic Development. *Annu. Rev. Energ. Environ.* 1996, *21*, 497–530.
(363) Bailis, R.; Ezzati, M. and Kammen, D.M. Mortality and Greenhouse Gas Impacts of Biomass and Petroleum Energy Futures in Africa. *Science.* 2005, *308*, 98103.

INDEX

Printed and bound by CPI Group (UK) Ltd, Croydon, CR0 4YY

17/10/2024

01775689-0010